Amazon Echo Guide

Newbie to Expert in 1 Hour!

by Tom Edwards & Jenna Edwards

Copyright © 2018 by Tom Edwards & Jenna Edwards – All rights reserved.

AMAZON ECHO DOT & ALEXA are trademarks of Amazon Technologies Inc. All other copyrights & trademarks are the properties of their respective owners. Reproduction of any part of this eBook without the permission of the copyright owners is illegal – the only exception to this is the inclusion of brief quotes in a review of the work. Any request for permission should be directed to ReachMe@Lyntons.com.

001801031

Other Books By Tom & Jenna Edwards

Amazon Echo User Guide - Newbie to Expert in 1 Hour!

Amazon Echo Dot User Guide - Newbie to Expert in 1 Hour!

Amazon Echo Show User Guide - Newbie to Expert in 1 Hour!

Fire HD 7 User Guide - Newbie to Expert in 2 Hours!

Fire HD 8 & 10 User Guide - Newbie to Expert in 2 Hours!

Fire TV Stick User Guide - Newbie to Expert in 1 Hour!

Amazon Fire TV User Guide - Newbie to Expert in 1 Hour!

Before We Start - Important!

Throughout this book we recommend certain webpages that might be useful to you as an Echo Dot owner and user.

A lot of the webpages we recommend are on Amazon, and a typical Amazon web address might look something like this:

www.amazon.com/dp/B00DBYBNEE?_encoding=UTF8&camp=1789&creative=9325&linkCode=pf4&linkId=UPTTB3CK67NSPERY&primeCampaignId=prime_assoc_ft

The link above takes you to the Amazon Prime sign-up page, but that's a lot of characters for you to type! So we've used what they call a Link Prettifier to make the links shorter and easier to use.

In this example the shortened link is - **www.lyntons.com/USPrime** - if you type that into your browser, it will save you lots of typing time, but still take you straight to Amazon Prime....

Contents

Introduction   1

   Do You Need This Book?   1
   How To Use This Book   2

1. What is Amazon's Echo Dot?   4

   Here Are the Echo Dot's Technical Specs   8
   How Does the Dot Compare to the Rest of the Echo Family?   8

2. Setting Up Your Amazon Echo Dot   10

   What's in the Box   10
   Setting up the Main Unit   10
   Accessing the Alexa App   11
   Connecting to Wi-Fi   12
   Using the Alexa Voice Remote   14
   Using the Dot as a Bluetooth Speaker   15

3. Amazon Alexa App Basics   21

   The Home Page   22
   The Now Playing Page   26

4. Video   27

   Amazon Fire TV   27
   Dish TV   28

5. Music   31

   My Music Library   31
   Prime Music: What it Offers   33
   Spotify: What it Offers   38
   Pandora: What it Offers   40

iHeartRadio: What it Offers	44
TuneIn: What it Offers	46
SiriusXM: What it Offers	48

6. Books — 53

Audible: What it Offers	53
Kindle Books: What it Offers	55

7. Lists — 58

Shopping Lists	59
To-do Lists	61

8. Reminders & Alarms (and Timers) — 63

Reminders	63
Alarms	66
Timers	67

9. Routines — 69

10. Alexa Skills — 72

Enabling Alexa Skills	72
Tips for Picking Skills to Try and Managing Enabled Skills	75
Develop a Skill to Share	76

11. Smart Home — 77

Setting Up Your Smart Home Devices	77
Getting the Most Out of your Smart Home Devices	79
Troubleshooting Wi-Fi Issues with Smart Home Equipment	82
Top Smart Home Alexa Skills	83

12. Things to Try — 84

What's New	85

Echo Dot ... 86
Ask Questions ... 86
Calling and Voice Messaging ... 86
Check Your Calendar ... 87
Connect Bluetooth Devices ... 87
Control Music ... 87
Control Smart Home Devices ... 88
Control the Color of Smart Lights ... 88
Discover Music ... 88
Drop In - Alexa's Intercom System ... 88
Enable Skills ... 89
Find Local Businesses and Restaurants ... 89
Find Traffic Information ... 89
Fun and Games ... 90
Get Weather Updates ... 90
Go to the Movies ... 90
Hear the News ... 90
Keep Up with Your Sports Teams ... 91
Listen to Audible Audiobooks ... 91
Listen to Amazon Music ... 92
Listen to Kindle Books ... 92
Listen to Podcasts and Radio ... 92
Shopping ... 93
Set Alarms/Set Reminders/Set Timers ... 93
To-dos and Shopping Lists ... 93
Use These Phrases Anytime ... 94
More Alexa Products to Try ... 94

13. Settings ... 95

Devices ... 95

Accounts ... 98
General ... 104

14. Help & Feedback ... 105

The Alexa App User Guide ... 105
Alexa Device Support ... 105
Calling and Messaging ... 105
Drop In ... 107
Music, Video and Book ... 108
News, Weather & Traffic ... 109
Smart Home ... 109
Shopping ... 110
Alexa Quick Fixes ... 111
Contact Us ... 112

INTRODUCTION

Welcome

Welcome and thank you for buying the **Amazon Echo Dot User Guide: Newbie to Expert in 1 Hour!**, a comprehensive introduction and companion guide to the exciting possibilities that this all-new 2nd generation Echo Bluetooth speaker and personal assistant has to offer.

Do You Need This Book?

We want to be clear from the very start - if you consider yourself tech savvy, e.g. the kind of electronics user that intuitively knows their way around any new device or is happy Googling for answers **then you probably don't need this book**.

We are comfortable admitting that you can probably find most of the information in this book somewhere on Amazon's help pages or on the Internet - if, that is, you are willing to spend the time to find it!

And that's the point... this Echo Dot book is a time saving manual primarily written for those new to streaming media devices, Bluetooth devices and tech that works in tandem with your PC or mobile device.

If you were surprised or dismayed to find how little information comes in the box with your Dot and prefer to have to hand, like so many users, a comprehensive, straightforward, step by step Amazon Dot guide, to finding your way around your new device, **then this book is for you**.

Furthermore, as mentioned above, the Echo Dot and other Echo/Alexa enabled devices are still fairly new in the big scheme of things and there will be new features, skills and services, not to mention

Echo Dot tips and tricks, appearing constantly over the coming months.

We will be updating this Echo Dot manual as these developments unfold, making it an invaluable resource for even the tech savvy.

Even if you are buying the first edition of this Amazon Echo Dot instruction book, never fear, you too can keep up to speed with all the new Echo Dot updates by signing up to our free email newsletter here - **www.lyntons.com/updates** - so you'll never miss a thing.

How To Use This Book

Feel free to dip in and out of different chapters, but we would suggest reading the whole book from start to finish to get a clear overview of all the information contained.

We have purposely kept this book short, sweet and to the point so that you can consume it in an hour and get straight on with enjoying your Echo Dot.

This Echo Dot user manual aims to answer any questions you might have and offer Amazon Echo Dot information including:

- What is the Echo Dot and how does it work?
- What does the Echo Dot do?
- How to setup your Dot
- How to setup Alexa
- How to manage your Amazon Echo Dot account
- How the Echo Dot and Alexa work together
- Echo Dot tips
- Echo Dot specifications
- Echo Dot settings
- And a general Amazon Echo Dot review

This Echo Dot tutorial will also look closely at Dot features including:

- The Amazon Alexa voice remote (including Echo Dot voice commands)
- The Amazon Alexa app
- Echo Dot extras
- The Dot shopping list and to do list
- Dot radio, music and news
- How to use Alexa Skills
- Echo Dot smart devices (including home hubs and lights)
- Dot accessories
- Plus much, much more....

And for further Echo Dot customer support we have links to

- Amazon Echo Dot customer service
- Dot discussion forums
- Echo Dot feedback
- Dot quick start guide
- Dot videos
- Dot Help Pages, help desk and community

As we will be updating this book on a regular basis we would love to get your feedback, so if there is a feature that you find confusing or something else that you feel we've missed then please let us know by emailing us at ReachMe@Lyntons.com. Thank you!

So without further ado let's begin...

1. WHAT IS AMAZON'S ECHO DOT?

The Echo Dot is a puck sized personal assistant that is becoming more useful all the time. Amazon is consistently developing new capabilities and updating your Dot via its Wi-Fi connection.

While you might first think that the Dot is little more than a technological novelty that can do a few cool things, it has become a valued asset in the time we've been using it.

We turn to the Dot throughout the day for news, weather, music and general information.

It assists us in many ways with time management and scheduling, home automation, online purchasing, communication and social media, with more uses being added on a regular basis.

This guide will help you get the most fun and functionality from your Echo Dot. We want to set your mind at ease from the very beginning:

Setting up the Dot is simple. From there, each step in tailoring this smart device to fit your lifestyle is easy and takes just a few minutes.

To personalize the interaction, Amazon has given the device the name "***Alexa***," and that is the wake word used to activate it.

Talking with the device really is like a conversation due to its pleasant and fluid voice, and "she" will soon be assisting you in lots of wonderful ways.

By the way, if you find it odd to address an inanimate object using a woman's name, you can use one of these alternate wake words, "***Amazon***", "***Echo***" or "***Computer***".

We will explain how to set your preferred wake word later. We've tried each option and have finally stuck with Alexa. It seems perfectly natural.

The Echo Dot itself is just a speaker with microphones inside. To bring it to life you will need to get familiar with the Alexa app, a vital piece of the Amazon Echo system.

In short, the App is a piece of software easily downloaded to your smart phone, tablet or accessed via a browser on your PC or Mac, and it works in tandem with the Echo Dot to allow you to get the most from its capabilities.

The Alexa app is, essentially, the remote control center for the Echo Dot and this guide will take you through every aspect of the app.

Voice recognition software is the key to your communication with the Dot. The device is programmed to understand US and UK English and will comprehend most users without a problem.

We've had some fun speaking to the device using exaggerated intonations and inflections including a variety of badly performed accents, and Alexa has understood us remarkably well.

AMAZON ECHO DOT USER GUIDE

Now, let's introduce this attractive cylinder and how it works. You'll notice four buttons on top of the Dot. The first is the Microphone On/Off Button.

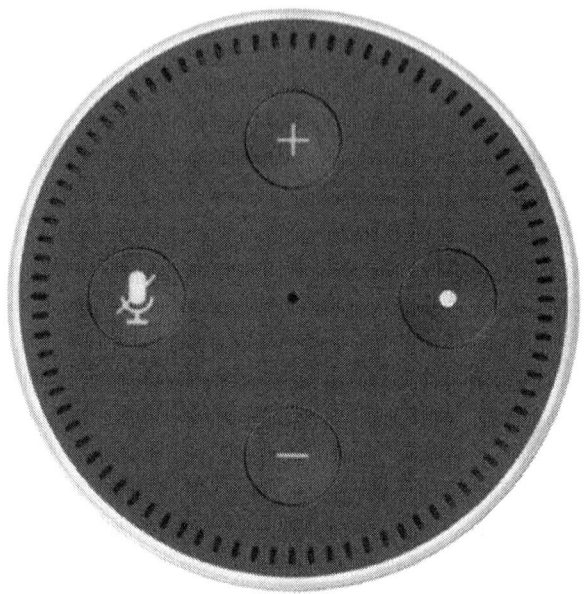

When they are switched on, the microphones are listening for you to speak the wake word.

When turned off, the mics can't hear you. It's that simple.

You might turn off the microphones if you're using the Alexa voice remote, which is sold separately.

We'll explain how easy it is to communicate with the Dot using the remote in **Chapter 2**.

The second button on top, the one with the single raised dot, is called the *Action Button*.

It can be used to wake the Echo in lieu of saying the wake word, turn off a timer or alarm sound and enter Wi-Fi setup mode.

The third and fourth buttons, with the + and - symbols, are your Volume Buttons. You can also give voice commands to alter volume such as, "*Alexa, louder, softer, volume 5*".

A light ring circles the top of the device and serves as an indicator of the Dot's status. Here's a list of the most common light ring colors and what they mean:

- **Orange spinning light:** Your Dot is connecting to the internet
- **Solid blue with spinning cyan lights:** The Dot is starting up when first plugged in.
- **All lights off:** The Dot is waiting for you to wake it using the wake word or the action button, or it is unplugged.
- **Solid blue light except for a small cyan section:** The Dot is listening to you or replying; the cyan shows which of the seven microphones is picking up your voice.
- **Flashing blue and cyan lights:** The Dot is processing your request.
- **Alternating blue and cyan lights:** The device is giving a longer answer.
- **Solid red light:** The microphones on the Dot have been turned off - To turn them on, press the Microphone button on top of the unit.
- **White light:** The Echo's volume is being adjusted.
- **Oscillating violet light:** An error has occurred during Wi-Fi setup.

There are some other color combinations to reveal but we'll hold back on them for now!

The Echo Dot is equipped internally with seven microphones to pick up your voice from any direction.

They are turned on using the Microphone button. When it is on, the device starts listening and is ready to assist you when you say the wake word or press the action button.

Here Are the Echo Dot's Technical Specs

Dimensions: This cylindrical speaker is 1.3 inches tall with a diameter of just 3.3 inches.

Device Type: Wired speaker

Connections: Bluetooth and Wi-Fi

Speakers: 1 - Tweeter, 0.6 inches

Available fabrics and finishes: Charcoal Fabric, Indigo Fabric, Sandstone Fabric, Merlot Leather, Midnight Leather, Saddle Tan Leather

How Does the Dot Compare to the Rest of the Echo Family?

We wouldn't blame you if you felt a bit overwhelmed by the number of Echo devices recently released by Amazon. Things were pretty straight forward when you only had the Echo, Echo Dot and Echo Tap (the portable Echo) to choose from.

Now, as of October 2017, Amazon has now added the Echo Plus, Echo Show, Echo Spot and Echo Look into the mix, all at different price points and each with different features. The good news is that at least one of these Echo devices will be right for you and at the right price. The slightly less good news is that you will have to work out which one to choose!

We're assuming (always problematic!) you've bought this book to help you with your new Echo Dot, but let's take a moment now to look briefly at the other options in case there's a further device that you think would work well in your household.

Echo Dot: So you've bought the Echo Dot which is basically a mini Echo. It only has a tiny speaker so isn't the best way to listen to music but it can be plugged into any other sound system in your home and it has all the Alexa features and functionality. If you think of the Echo as the main device for your living room then the Dot is the perfect addition for a bedroom. At just $50 it's perfect if you want Alexa in every room of your house.

All-new Echo: The second generation Echo has had a visual overhaul, we think it looks 100% more attractive, but retained all it's great features and at a lower price. If you want to play music through your Echo device then this is the best device as it will fill a large room with music comfortably. Our approach is to have an Echo in our main living area and a couple of Dots in the master bedroom and home office.

Echo Plus: The all new Echo Plus is really more like the 1st generation Echo but with an integrated smart home hub. It's the same size as the original Echo, has an improved speaker system and, as mentioned, a built in smart home hub. The integrated hub makes connecting to smart home devices easier for sure, but it's not essential.

We already use a Wink hub to get our smart home devices online so adding the Echo Plus to our home doesn't really make any sense, we're already set up. If you're new to smart home devices and want to jump in, then the Echo Plus would be a great device to buy. If you feel lukewarm about the smart home revolution, then we would suggest sticking with the standard Echo for now.

Echo Show and Echo Spot: These two new devices are basically the Echo and the Echo Dot but with screens! It's as simple as that. If you prefer to see rather than just hear the information and entertainment that Alexa provides then these two are for you. We have both and like both, but you need to decide if you need another screen in your life. The screens on both devices are small so it's not as if you'll be likely to watch a movie on them when you can watch on a TV or tablet.

The Echo Show is, we think, the top of the line Echo device but with a price to match. If you want the best device then the Show is for you. The Echo Spot is a very tempting addition to your Echo household as at $129.99 (at the time of writing) it doesn't feel like an extravagance. If you like the visual aspect of having an Echo with a screen then the Spot is your budget friendly option.

2. SETTING UP YOUR AMAZON ECHO DOT

The setup was super-quick, and much of it happened automatically with little or no input from us. Here are the details.

What's in the Box

- Amazon Echo Dot
- Power Adapter
- Quick Start Guide
- Things to Try Card

The Alexa voice remote, which you might find very handy, is sold separately by Amazon for $29.99. However, it is not needed in order to use your Dot.

Setting up the Main Unit

You can locate the Dot in any room of your house, but its small speaker makes it less useful for large spaces, unless you plan to connect it to a larger sound system (more on that later). So in our house the Dot lives on the bedside table, it's perfect there for checking news, weather or traffic info and listening to audio books or podcasts.

When it is plugged into the new location, it reconnects to our Wi-Fi system by itself in 30 to 75 seconds.

Amazon recommends that the device be placed at least eight inches away from walls and windows to ensure clear communication, perhaps to avoid confusing the Echo with an echo!

We've tried it against a wall, and it worked fine. Just keep that tip in the back of your mind in case your Dot parked on a window sill or near a wall is having communication problems.

Plug the power adapter into the Echo Dot and into a 110-120 volt outlet. The light ring will turn blue to acknowledge that it is powered and then turn orange. At this point, Alexa will greet you with "Hello".

Simple to Set Up & Use

1. Plug in Echo Dot
2. Connect to the Internet with the Alexa App
3. Just ask for music, weather, news and more

Now, it's time to get familiar with the Amazon Alexa App.

Accessing the Alexa App

Setting up the Dot takes just a few minutes using the quick- start guide included with your Echo device, and it worked flawlessly when we tried it. First up you need to decide which device you're going to use to access the Alexa app prior to setting up your Echo Dot.

Our preferred method is to access the app via a browser (Chrome/Safari/Firefox etc) because of the larger screen size, so if you want to try this method then open a browser window on your pc or mac and enter this address **alexa.amazon.com**

If you would prefer to access the Alexa app via your mobile phone or tablet then go to the app store of your device (e.g. Google Play, Apple App Store or Amazon App Store), search for "Alexa app" and then download and open the app.

Regardless of which method you choose you will need to sign into the app using your Amazon username and password.

AMAZON ECHO DOT USER GUIDE

Connecting to Wi-Fi

Once logged in, the Alexa App walks you through the short process of connecting the Echo Dot to a Wi-Fi network. Have your Wi-Fi password available. The password is usually located on the Wi-Fi router. If it's not there, contact the company that provides your Internet service.

Note: the Dot does not support enterprise or ad-hoc/peer-to-peer networks, though it is unlikely that your network is one of these types.

If this is your very first time accessing the Alexa app you will automatically be taken through the setup process. You will be asked to choose the device you are trying to set up and then you will be walked through the steps to link your Dot to your Wi-Fi network.

The light ring on your Dot should be a slow spinning orange at this point, which indicates it is trying to connect to your Wi-Fi network. If you don't see an orange light press and hold the ***Action Button*** on the top of the Echo Dot (the one with the raised spot) until the orange light starts spinning.

At this point the Alexa app will prompt you to choose a Wi-Fi network or it may direct you to go to the Wi-Fi network settings on your pc, depending on what device you are using to access the Alexa app. Just follow the instructions and enter your Wi-Fi password if requested. And all things being well that should be that.

If this is not your first time accessing the Alexa app and you're setting up a new Dot alongside an existing Alexa enabled device then simply plug in your Dot, wait for the orange spinning light, log into your Alexa app and find the Settings tab in the navigation panel on the left hand side.

Select **Settings > Set up a new device > Echo Dot**. By the way, this sequence means you should:

1. Select **Settings** from the options on the Alexa App homepage

2. Select **Set up a new device** from the options that appear

3. Select **Echo Dot** from the set of device options you're given

Now, as described above, follow the setup process and connect to your Wi-Fi network.

If your network isn't on the list, scroll down to select **Re-scan** to search again or choose **Add a Network** and follow the instructions that appear.

When we did this, we didn't strictly time the process. It seemed to take more than a minute and less than three for the Dot to connect to the network. When it did connect, a confirmation message appeared on the **Set up a new Echo** page of the app....so be patient.

You're now ready to return to the app *Home* page, which you'll find at the top left of the screen, to begin using your Dot.

Start by addressing the Dot with the default wake word "*Alexa*". The wake word can be changed to "*Amazon*" , "*Echo*" or "*Computer*" by selecting **Settings > [Your Name]'s Echo Dot > Wake Word > Amazon > Save**.

Where it says *[Your Name]'s Echo Dot* above, we mean the name you change your Dot to during the "**Set up a new device**" process. We named our Dot '**Jenna's Echo Dot**', so in our case it would be **Settings> Jenna's Echo Dot > Wake Word > Amazon > Save**.

13

If we ever want to change the name there's an option to change it, just go to **Settings > Jenna's Echo Dot > Device name > Edit**

Using the Alexa Voice Remote

The Alexa Voice Remote is optional and costs about $30. Its value is in allowing you to make requests from a distance when there is noise interference between you and the device or you don't want to speak loudly.

Requests you make are digitized and sent to the Echo Dot. We have a remote, and it works fine, but after collecting dust for a few months, it was put in the gadget drawer and is rarely used.

There are two concerns many users have about the Alexa Voice Remote.

First, its voice recognition and communication capabilities are mediocre, so more commands/requests than usual will be confused. Secondly, in the past, they have simply stopped working at an unacceptably high rate. Amazon has stopped production of the remotes several times to work out the bugs, so read a handful of recent negative reviews to learn what the current issues are before you buy one.

One other consideration, as mentioned you can only pair one Bluetooth device to your Dot at a time, so if you buy and pair the remote you won't be able to pair your mobile phone and the Dot at the same time.

If you plan to purchase one, make sure it is the Alexa Voice Remote and **not** an Amazon Fire TV remote. **They are not interchangeable**.

If you have the Fire TV Stick with Alexa Voice Remote (currently about $40), that is a combination remote that will save you money over buying two remotes.

If you purchase one, you'll need to pair it via the Alexa app:

To start, make sure the two AAA batteries have been inserted and that they're oriented properly. In most cases, the remote will automatically pair with the Dot. If it does not:

Go to the Alexa App

Within *Settings*, select *Your Echo Dot*

Then select *Pair device remote*

The Dot will search for the remote and ask you to press and hold the play button on the remote for 5 seconds to complete pairing

If you ever want to remove the remote choose the *Forget Remote* option in **Settings** under *Your Echo Dot*.

Using the Dot as a Bluetooth Speaker

As you probably already realize, the Echo Dot is a paired down version of the Amazon Echo. While the Dot does contain a speaker it is nowhere as sophisticated or as powerful as the speaker on the larger Amazon Echo.

That's not to say the Echo Dot's speaker is useless, we've found that it's perfectly good enough for everything except high quality music playback.

So if you're getting spoken information from Alexa (news/traffic/general knowledge), listening to an audio book or simply want to have some quiet music playing in the background then the Dot's small internal speaker is sufficient.

But if you want your Echo Dot to really sing then you can connect it to a separate external speaker, either via Bluetooth or a wired connection, and your sound quality will be just as good, if not better depending on your external speaker, than using the original Amazon Echo.

So far so good!

However the way you use your Dot will affect some of the things that you can do and how you should set things up.

So let's now look at three different sound set up scenarios and see which is best for you.

Using Your Echo Dot's Internal Speaker

As you work your way through this guide you will discover all the different features that the Echo Dot offers. There are lots of them and they are the main reason you bought your Dot in the first place!

That said you can also use your Dot simply as a standalone Bluetooth speaker, all be it a not very powerful one. But why would you want to do this? Well the simple answer is that we all have mobile phones or tablets or computers that we spend a lot of time on and these devices also have lots of audio content, like music, podcasts, audiobooks.

We listen to a lot of podcasts and we do that almost exclusively through our mobile phones.

Therefore for us it's extremely useful, when we're at home, to stream these podcasts through an external speaker… an external speaker like the one in the Echo Dot!

So, if that's something you would like to do this is how you go about it:

To stream content from any other Bluetooth enabled device you may have using the Echo Dot as your speaker, it must first be paired to the device containing the content you want to hear.

This is done by finding and accessing the Bluetooth settings of the phone or tablet you want to pair your Echo Dot with.

Obviously the location of the Bluetooth settings will vary from device to device, they are typically found in the general Settings menu of your device. The first time you pair a mobile device we recommend you use the Alexa app via your PC or Mac and follow these steps:

Make sure Bluetooth is turned on in the Settings section of your mobile phone or tablet

Now go to into your Alexa app and visit ***Settings > Your Echo Dot > Bluetooth > Pair a New Device***

When you click/tap ***Pair a New Device*** a page will open and search for "Available speakers" to connect to. **Tip:** You have already turned on Bluetooth on your mobile, but also make sure that the phone/tablet is active, that's to say not in sleep mode.

Your phone or tablet should now appear as an Available speaker so go ahead and click on it.

Your mobile device will probably flash up a message asking you to confirm that you want to pair it with your Dot, so go ahead and confirm.

A single chime will sound to indicate the device has been paired, and Alexa will say "Now connected to [your device]"

From now on you can unpair or pair this mobile device by simply saying "***Alexa disconnect***" or "***Alexa, connect***"

Repeat this process for every new device you want to pair with Alexa.

Also be aware that you can only pair one Bluetooth device at a time. If you've paired several devices to your Dot, it will connect to the most recently paired when you make the request. To change from one mobile device to another you have to return to **Settings > Your Echo Dot > Bluetooth** and manually disconnect from the current connected device before reconnecting to a different device.

When a Bluetooth device is paired to your Dot, you'll enjoy hands-free control with commands such as, *"Alexa...*

- *Play"*
- *Pause"*
- *Previous"*
- *Next"*
- *Stop"*
- *Resume"*
- *Restart"*

We found through experience, though the information is on Amazon too, that if you try to request specific songs, albums or artists from the music stored on your phone or other Bluetooth device, the Dot will disconnect from the device and search for the music in your Amazon Music library.

So you can only use these voice commands to control content that is already playing.

To reconnect to the Bluetooth device, simply say, "***Alexa, connect***", and press play on the device to continue listening.

It should be stressed that before you can use these commands you need to start playing something via your paired device, play an album in your iTunes app for example or start playing an audiobook or a podcast.

Using Your Echo Dot With a Wired Connection To An External Speaker

For this scenario let us now imagine that you want exactly the same set up, as we've just discussed in the previous scenario, but this time you want excellent audio quality.

You want all the features of the Echo Dot plus you want to be able to stream audio content from your phone or tablet and you want that audio to sound great.

In this case you follow exactly the same instructions as above and pair your phone or tablet with the Dot, but you also need to connect your Echo Dot to an external speaker via a 3.5mm audio cable. You will need to buy this audio cable separately.

Just so you know, when your Echo Dot is connected to a speaker via the audio cable the external speaker will override (shut down) the internal speaker of your Dot... even when the external speaker is switched off.

So when the two are connected via an audio cable make sure the external speaker is turned on before trying to interact with your Dot.

Connecting Your Echo Dot to an External Speaker Via Bluetooth

If you want the convenience of connecting your Dot wirelessly to a Bluetooth enabled speaker then it's pretty simple. Yet again follow the pairing instructions we gave you earlier.

The only difference will be that to turn on Bluetooth connectivity on your external speaker you will need to press a button on the speaker. Where that button is will depend on the Bluetooth speaker you are using.

Connecting wirelessly means you can place your speaker anywhere in the room, it doesn't have to be near/wired up to your Echo Dot.

However the downside is that you can no longer stream content from any other device, like a mobile phone, while using your Echo Dot.

Basically your Echo Dot can only pair with one Bluetooth device at a time so if it's paired with your speaker it can't then pair with your phone at the same time.

If you are someone who is really invested in Amazon and all their technology this may well not be a problem. If you already listen to all your music using Amazon Music or if you're happy to listen to your podcasts via TuneIn which is supported by the Alexa App (more on that later) then you won't care about streaming audio from your phone because you will already have access to all the content that you like via your Dot and the Alexa app.

So let's now move on and look more closely at the Alexa app, what it can do and what content and great features you can access through it.

AMAZON ECHO DOT USER GUIDE

3. AMAZON ALEXA APP BASICS

The smart Echo Dot Bluetooth speaker and the Amazon Alexa App work together as a team to create a really cool personal assistant that we expect to become more competent as time goes by, via Amazon updates and the addition of new third-party partners. Together, the app and device form a system called Amazon Echo. This chapter is an overview of the Alexa App and how to use it. In subsequent chapters we will go into each section of the Alexa App in detail.

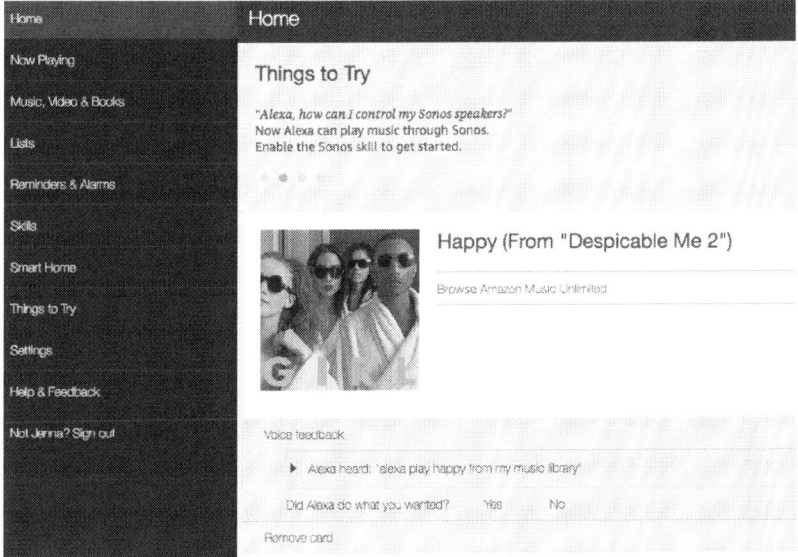

We first explored the app on our laptop because navigation is simpler and the large screen is easier to see. Once we were familiar with the app, working with it from a cell phone is a snap.

Throughout this guide we offer step by step instructions and we recommend that you follow these instructions via the Alexa app on your computer at **alexa.amazon.com** first before trying the same sequence on a mobile device.

When you first log into the Alexa App on your computer you will arrive at the **Home** page. You will see a menu of options to the left and the contents of the Home section to your right.

21

AMAZON ECHO DOT USER GUIDE

On mobile devices, because the screen is smaller, you won't see the side menu of options, instead, top left you will see the menu icon, three horizontal lines, which you can tap to reveal the menu of options.

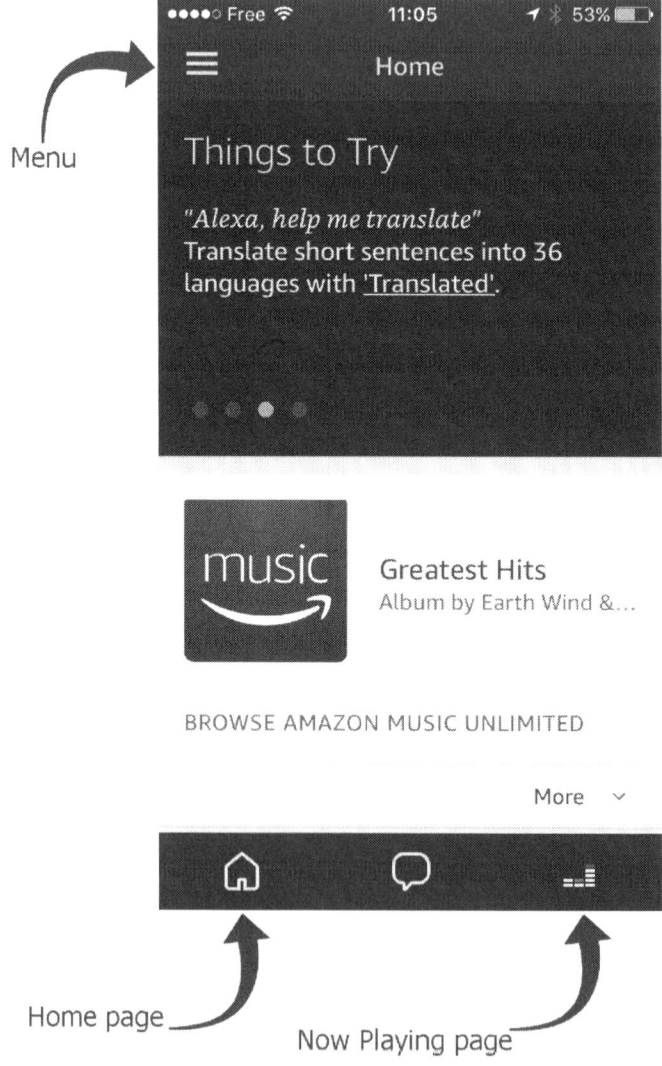

The Home Page

On a computer **Home** is the first item in the left menu. On a mobile device you can always reach the Home page by taping the little house icon at the bottom left of your screen.

The Home page itself is split into two sections. Right at the top you will always see a selection of **Things to Try**. This is where Amazon promotes the latest fun and interesting Skills (**see Chapter 10**) new to Alexa.

There will typically be three to five new voice commands that you can try. Frankly, many (most) of these things to try are fairly silly. Occasionally our interest is peaked but if you're like us you might well find yourself ignoring these snippets once the novelty of owning you Dot has worn off. Please get in touch if you disagree!

Below **Things to Try** you will see your dialog cards. One card is created for each vocal request or command that you give to Alexa. So if you haven't yet asked Alexa for anything why not choose one of the Things to Try now. After you've finished you should see a card appear with information of the request you made.

To the right of each card you will see the word **More** and a down arrow, clicking or tapping here will give you the option to provide feedback to improve your Dot's voice recognition.

What Alexa heard is listed with the question, "**Did Alexa do what you wanted?**" Answer **Yes** or **No**. You've also got the option to remove the card.

At first, we gave feedback and deleted a few cards to reduce clutter, but we mostly forgot about it pretty quickly.

Now the only time we choose the down arrow and give feedback is if the Dot did not hear us correctly, a rare occurrence. We also occasionally scroll back through the cards to locate a song/playlist/station we enjoyed but forgot to get the name of.

Depending on the type of request that produced the card, you'll have more options. For example, if you've requested Dot to play something from your Amazon Music Library, the card includes options to **Search the Amazon Digital Music Store** for similar music, **Search Amazon Prime** for free music and **Browse your library** for similar choices.

Cards for information requests will include the option to **Search Bing** for the topic.

If you choose the **Learn More** option in blue on any card or at the top of the list of cards, the Alexa FAQs appear.

You can manage the history of your dialog with Amazon Dot in much the same way as you manage your computer's history. Go to **Settings > History** where you'll find the complete list of requests and commands.

We occasionally scroll down the list to find a station we were playing or retrieve an answer to a question we asked. Provide feedback or delete any request by selecting the down arrow.

If you prefer to delete all voice recordings you actually have to visit your Amazon account on the main Amazon website and find the **Manage Your Content and Devices** page of your account. Here is a direct link www.amazon.com/mycd

To find this page manually, go to **www.amazon.com** and log in. Towards the top of the page to the right you will see **Account and Lists**.

Hovering your mouse over this will reveal a large drop down menu and about half way down you will see **Your Content and Devices**, click it.

This is an incredibly useful and important page of your Amazon account for anything to do with the digital content you buy from Amazon, the electronic devices you have registered with Amazon and all associated settings.

If we're ever unsure how to delete or change something in relation to our content and devices we inevitably find the answer on this page.

So returning to the question of how to delete all your voice recordings...once you are on the **Manage Your Content and Devices** page select the **Your Devices** tab. Find your Dot and click the small box to the left. A popup window will appear with some options including **Manage voice recordings**. Keep in mind that Amazon Dot has been learning how you speak to give more consistently accurate answers. If you delete your voice recordings, you might find that the Dot's ability to understand you is reduced.

You'll have plenty of time to come back to the Home page later once you start using your Dot in earnest so we suggest you revisit the above information at a later date. For now let's move on and run through a brief description of the other menu options

The Amazon Alexa App menu on the left gives these choices:

- **Home:** As discussed, when Home is selected you will see some Things to Try and your list of cards.
- **Now Playing:** Shows you in detail exactly what you are currently listening to.
- **Music, Video & Books:** This is where you can find info on your Amazon Music Library, Prime Music subscription, Spotify, Pandora, iHeartRadio, TuneIn, Audible and more.
- **Lists:** Create shopping or To-do lists manually or by voice
- **Reminders & Alarms:** Here you can setup and manage multiple reminders, alarms and timers.
- **Routines:** A feature that allows you to activate several functions with one voice command
- **Skills:** Discover all the third part apps that add further functions and features to your Dot
- **Smart Home:** This is where you connect your smart home devices to your Dot
- **Things to Try:** Not to be confused with the small selection of novelty items previously discussed on your Home page. This list is a great resource for keeping up to date with developments and trying out features.
- **Settings:** We'll refer to Settings throughout this guide to help you manage your Amazon Dot and account. And we have a chapter dedicated to this important section.
- **Help & Feedback:** Valuable guides can be found here to help you with every aspect of Alexa and your Dot

The Now Playing Page

The second item in the left side menu is the **Now Playing** page. As the name suggests this page of the Alexa App provides further information about the audio media (music, podcasts, audio books etc.) that you are currently listening to on your Dot.

When you're not listening to anything via your Dot there's not much to see on this page. In **Chapter 5** we'll explain how to get setup and start playing music and other audio on your Dot. So we will return to and talk more about the Now Playing page shortly… once you're actually playing something!

Now that you've been introduced to your new personal assistant, it's time to get to know Alexa more thoroughly and find out what her true capabilities are.

We've created the following chapters from our experience, and it's intended to be a guide you can refer to in the days ahead to maximize the fun and usefulness of your Amazon Dot and to minimize the hassle. Let's start with the Dot's newest feature…Video!

4. VIDEO

The next tab on the menu is *Music, Video & Books*. Covering all three would make for a long chapter and would also make it difficult for readers of this guide to locate the information they wish to review. So, we'll take these entertainment options one at a time.

While Music is listed first in the tab title, Video options are shown first on the app. To maintain our systematic approach, let's start at the top of the page and work down over the next three chapters:

Chapter 4: Video

Chapter 5: Music

Chapter 6: Books

At this writing, your Dot works in conjunction with Amazon Fire TV and Dish TV. Amazon has said the company is developing relationships with other video services, so expect this category to expand.

Amazon Fire TV

Fire TV is Amazon's streaming media player for your TV, designed to improve your experience when accessing live content and streamed video, music and games. It's just like the Apple TV and Google's Chromecast. Some of the content is free but much of it requires a paid subscription. Fire TV equipment can be controlled with Alexa devices including the Dot.

Content: The stable of streaming and live TV subscription services is expanding rapidly and already includes Netflix, Hulu, Amazon, HBO Now, DirectTV Now, NBA, MLB, CNN, Crackle, ESPN, Showtime, Sling, Comedy Central, HGTV and AMC. You must have a subscription to these services for them to be available on Fire TV and controllable by your Dot.

Setup And Use Fire TV with your Dot:

- Choose *Fire TV* from the *Video* section of *Music, Video, & Books*
- Select *Link Your Alexa Device* to choose which Fire TV equipment in the list you want to control with your Dot and Alexa
- Follow the prompts on the app to complete setup

Once setup is complete, you access and control Fire TV content on your TV using your voice:

Request the media you want to see or hear from free and subscription services by asking things like *"Alexa, search for Tom Cruise"* or *"...search for action movies"* or *"....search for Mission Impossible"*. You can also open specific services by saying *"Alexa open HBO Now"* etc

Once you're watching it, rewind or fast-forward streamed content for the length of time you choose using your voice and asking *"Alexa rewind 5 minutes"* or *"...play next episode"*

Dish TV

Dish TV is one of the most popular satellite TV services available. Packages start at $49.99/month for the service. You must also lease the Hopper DVR for $10/month.

Equipment: Hopper DVR and any Echo device including the Dot.

Information: Currently, Dish TV is offering new subscribers a free Echo Dot, the scaled-down version of the Echo. All the details are at **Dish.com**.

Content: The company markets Dish as a cost-friendly alternative to cable that also provides 30 premium movie channels and more HD channels among its total 190 channels. Among the most popular channels are HBO, NFL Network, NFL RedZone, ESPN, Disney Channel, USA, ShowTime, Starz and Cinemax.

Setup and Use of Dish TV with your Dot: Make sure popup windows are enabled on your browser. You might be asked to enter your Dish TV and Amazon login information during setup, so have it with you. Here are step-by-step linking instructions:

- Ensure that you have the latest version of the Hopper DVR software/firmware by going to channel 9607 on Dish TV and choosing Software/Accessibility Update
- If an update is needed, restart the Hopper when it is complete
- Choose *Dish TV* from the Video section of *Music, Video, & Books*
- Select *Enable Skill*
- Follow the prompts on the app to complete setup, which will include turning on the Hopper DVR and:

    1. Going to *Menu*

    2. Selecting *Settings*

    3. Selecting *Amazon Alexa*

    4. Selecting *Get Code*

    5. Entering the Code on the Alexa app where asked

    6. Selecting *Activate*

- Go to *Settings* on the Alexa app, choose *Hopper* from the list of discoverable devices and choose *Link Devices*
- Review the list of suggested things to say to control Dish TV with your Dot such as "*Alexa, tune to ESPN*" and "*Pause Dish*." The use is very straightforward. Tell Alexa what you want, and you'll likely get it.

The connection should go smoothly. If you have issues, details for resolution might be found on the Dish TV Alexa integration page found here - **www.dish.com/AmazonAlexaIntegration**

Additional help on Quick Fixes for Alexa Video Skills is found on Amazon at - **www.lyntons.com/AVS**

Note that you connect to Fire TV by linking your Alexa device with your Fire TV account, but you connect to Dish TV by enabling the Skill. They are two ways of doing the same thing. Alexa skills - and it's growing range of diverse capabilities - are covered in the Skills section to follow.

5. MUSIC

We're music fans, and you probably are too. Amazon gets this, so it built the Echo devices to deliver real listening pleasure. Let's look at the top music and audio services supported by the Amazon Echo Dot.

In this chapter we will take you through each music page, explaining what each service provides, how to set them up and how to control them with your voice or the Alexa App.

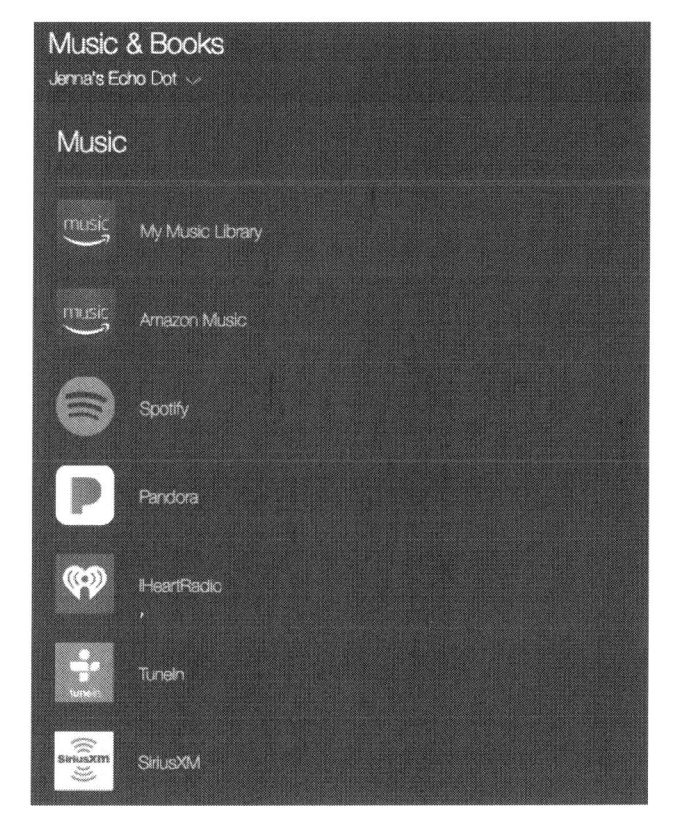

My Music Library

The first item under Music is My Music Library. Here you can access and play all the music that you have:

- Already purchased on **Amazon.com**

- Selected from your Prime Music or Music Unlimited subscriptions if you have either.

We will discuss **Prime Music** and **Music Unlimited** subscriptions in the next section of this chapter. And not much more needs to be said about music you have already purchased. Any music you've already bought from Amazon will automatically appear here in *My Music Library* and you just have to click or tap any track that you see here while using the Alexa App to start it playing.

And of course the whole point of Alexa is to operate it/her with your voice, so you can also start a track playing by saying "*Alexa play Can't Stop the Feeling by Justin Timberlake*" or the name of a track, album or artist that you have previously purchased.

Finding/viewing the music you would like to listen to in your **Music Library** is quite straightforward.

Search your Library: This search box is a quick way to find an Artist, Album or Song. The search results list will begin populating as you type.

Playlists, Artists, Albums, Songs, Genres: My Music Library is divided into these five categories. Select any of them, and all available options will be shown. The A through Z list that appears in each category, except Playlists, allows you to quickly search for something in the list, reducing the scrolling required.

Prime Music: What it Offers

There's a long list of benefits for using Amazon Prime including Prime Music. There's a 30-day free trial here - **www.lyntons.com/USPrime** - if you want to try it out before making a commitment.

With Prime Music you have unlimited ad-free access to more than two million songs in addition to complete albums and stations created for all major music genres.

If you're not interested in Amazon Prime then rest assured there are plenty of other free music options available that we discuss later in this chapter.

Amazon are now offering a further music subscription called Amazon Music Unlimited, for a monthly subscription fee you can listen to even more music via your Dot.

If you would like to know more visit this page - **www.lyntons.com/Amusic**.

Personally we find that the music options offered via our Prime membership is sufficient for our needs.

Exploring Your Amazon Music Account

Once you've joined Amazon Prime you can return to the Alexa App and click on *Music, Video & Books > Amazon Music* and start to explore your options.

It would be an understatement to say that there are a lot of musical options on the Prime Music homepage.

Frankly, there are so many options, and so many of them overlap, that it was initially confusing.

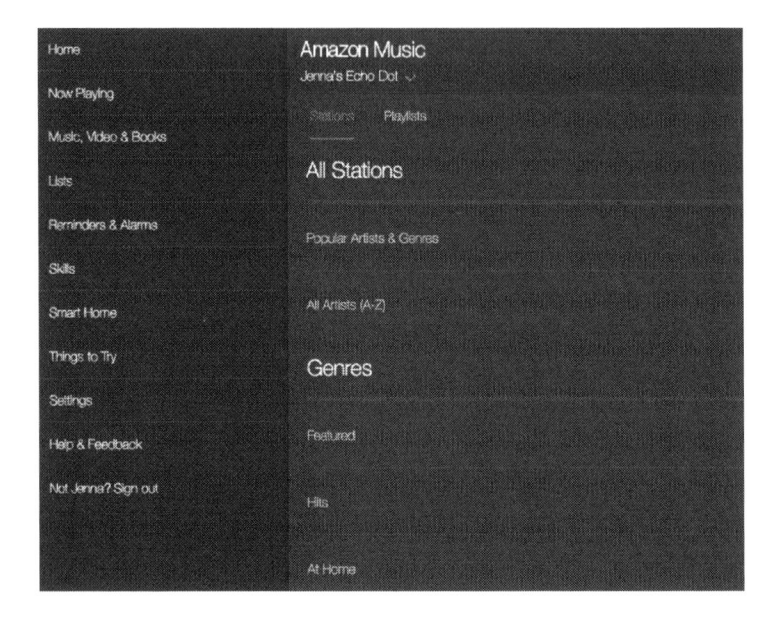

Think of it this way: The Amazon Prime music categories discussed below attempt to organize the entire catalog of modern music into logical groups which Amazon term as **Stations** and **Playlists**. There aren't hard and fast distinctions between these groups.

We quickly realized that there are just too many options to get a handle on easily so we adopted an adventurous, open-minded attitude.

We pick a station or playlist that sounds good to us, and wait to hear what songs are included in the collection.

We then note the station and return to it or avoid it based on our tastes.

Okay, with that perspective in mind, here is an overview of the musical groupings. Take a relaxed hour or so to explore what's available and get a feel for what directions you want to explore first.

Prime Stations

The dozens of stations cover all the musical styles we're familiar with, and we're music buffs, plus lots we're not acquainted with but look forward to exploring.

Here's something we didn't expect when we began to listen: Stations named for a specific artist play that artist as well as similar artists.

For example, the James Taylor station plays plenty of JT, of course, but also Christopher Cross, Joni Mitchell, Jim Croce, Gordon Lightfoot, Carole King, Jackson Browne, etc.

All Stations: The first thing you will see when you click on Stations within the Amazon Music page on the Alexa App is the All Stations heading and below that Popular Artists & Genres and All Artists.

The next heading is **Genres** which are divided into favorites such as Alternative, Blues, Children's Music, Classical, Classic Rock, Decades, Gospel and Pop. Each genre offers stations named for artists, decades and styles.

Prime Playlists

There are thousands of Playlists that give you a more specific slice of the types of music you like, and your options are incredibly diverse.

You can search Playlists by *Mood & Activities, Genres, Artists* and *Decades* to locate those that appeal to you.

Select "*Add Playlist to Library*" and the list will be transferred and stored there to be played on your Dot, phone or other device using the Alexa App and/or your voice.

When we clicked into *Mood & Activities* we didn't have to go further than *Happy & Upbeat* to find a whole bunch of playlists that we love!

See what we mean? Amazon Prime offers a torrent of music, and there are innumerable places to jump into the rushing waters. We think you'll enjoy the ride, as we do.

Top Tip: On the **Now Playing Page,** when any track is playing from a Prime Station or Playlist you can save it to your Music Library. Find the track to the right of the Now Playing page under ***Queue***, click the small down arrow and then click ***Add this song to your library***.

BUT what if you want the entire album that a track was taken from? Playlists or Stations with a selection of tracks, all from different albums, is great but sometimes you just want to hear one specific album from beginning to end. For the moment Amazon doesn't always make it easy to save a whole album via the Alexa App so here's our workaround.

Let's take the track "Happy" by Pharrell Williams which we came across in one of our Happy & Upbeat Prime Playlists. When the track is playing we can see on the Now Playing page of the Alexa App that the song is taken from the album G I R L.

Now we go to **Amazon.com** and type "*G I R L by Pharrell Williams*" into the search box and straight away we can see the album cover and details including a **Listen Now** tab indicating that this album is free for us to listen to as Prime members.

Clicking on the ***Listen Now*** tab instantly opens the Amazon Music app in a new browser window and the album starts playing through our pc speakers. Under the album title, as well as a pause button you will see a box saying "+ ***Add to My Music***". When you click that box the album is immediately added to My Music Library and now all I have to do to hear that album in its entirety is say "***Alexa play G-I-R-L by Pharrell Williams***" and…the album starts playing via my Dot!

Note: We've found that this whole process is a lot easier when listening to a Prime Station, as opposed to a Playlist.

When playing a Station go to the **Now Playing** page and under **Queue**, click the small down arrow and as well as *Add this song to your library* you should also see the option *View album in Prime Music*. Click or tap here to go directly to the album page on Amazon and then follow the instructions above to add it to your library.

Controlling Prime Music with your Dot and App

Using voice, it is as easy as saying, "*Alexa, play 50 Great Happy Songs*" to get the music started. The Amazon Dot will reply with "The Playlist 50 Great Happy Songs from Amazon Music," and a great selection of happy toe tapping music will fill the air through the Dot's speaker.

Once the chosen list is playing, Alexa will respond to commands that control the music. These include, "*Alexa:*

- *What's playing?"*
- *Turn it up or Softer"*
- *Volume 5"*
- *Skip or Next Song"*
- *Pause, Resume or Continue"*
- *Loop"*
- *Stop"*
- *Continue"*
- *Add this song"* (to your Amazon Music Library)

When using the Alexa App instead of your voice, simple taps or clicks are all that are needed to choose the music you want to play. When you've made your choice, a player appears with options for play, pause, skip ahead, go back or shuffle the music.

Spotify: What it Offers

This Swedish streaming service hooked up with Alexa after iHeartRadio and several others were onboard. In similar fashion, Spotify delivers music (30 million songs and counting) and podcasts to Dot users through searchable artists, albums, labels, genres and playlists.

While Spotify has both Free and Premium subscriptions, you'll need to go with Premium to listen to the service using Alexa. The current cost for Spotify Premium is $9.99, but a 30-day free trial is available that gives you ad-free streaming, unlimited skips and play any track features.

Of course you can use a Spotify free account with your Dot too. Remember that you can connect your mobile phone to your Dot via Bluetooth. When you do this whatever you play on your phone, including music playing in the Spotify app, will play through your Dot speaker. Not the perfect Alexa integrated solution, but certainly a budget workaround.

Setting Up Your Spotify Account

Go to **www.spotify.com**, click **Get Spotify Premium** and sign up. Once you've established a Spotify account, synching it to the Alexa App on your laptop or phone will take just a couple of minutes.

Go back into the Alexa App, go to **Music & Books > Spotify** and click **Link your account.** You'll then be taken back to the Spotify website to authorize the connection between Alexa and Spotify. It's that simple.

Back on the Spotify page of your Alexa App you'll be asked whether you want to make Spotify your default music library. Hold fire on that for a moment as we're about to explain how to operate Spotify with your Dot next.

Controlling Spotify with Your Voice and App

Using voice commands: Speaking directly to the Dot or using the Alexa voice remote, ask, "***Alexa, play Twenty One Pilots on Spotify***." It's important to include "***on Spotify***" or Alexa will try to select music from your Amazon Music library first. That is unless you make Spotify your default music service!

If you believe that you will be using Spotify to play the majority of your music on the Dot then setting Spotify as your default music library makes sense.

With this service set as default you can ask for music without specifying Spotify, Alexa will automatically look to play a music choice via Spotify rather than anywhere else. Of course if you choose this setting you will now have to specify Amazon Music by name if you want the Dot to play something from that library.

To set Spotify as your default music library go to ***Settings*** in the Alexa App then scroll down and click the ***Choose default music services*** box. Then choose ***Spotify*** under ***Default music library*** and finally click ***Done***

When requesting music from Spotify you can ask for a specific song name, playlist name, genre, artist, composer or Discover Weekly to hear what's new. For example, say, "Alexa, play songs by Carrie Underwood on Spotify," or "Alexa, play Work by Rihanna on Spotify."

Command options are what you'd expect: Pause, resume, stop, mute, previous, next, shuffle, skip this song, volume 1-10, volume up/down, etc. Remember to use your wake word.

You can get answers to questions like, "*Alexa, what song is playing?*" and "*who is this artist?*"

Using the Alexa App: Use the Now Playing page to view the current Spotify music you're listening to and use standard command options such as play, pause, skip, previous, next and shuffle.

Using the Spotify App on your phone: If you prefer the interface of the Spotify app you can use it to stream music directly to your Dot. With the Spotify app navigate to ***Devices*** and select ***Amazon Echo Dot*** from the device list. From there, voice commands will work.

Unlinking Spotify and Amazon Accounts

If you don't want the accounts linked:

- Open the Alexa App
- Browse the **Menu**, and choose **Settings**
- Choose ***Music & Media***
- Choose ***Spotify*** from the list
- Choose ***Unlink account from Alexa***
- Choose the ***Unlink*** option

Pandora: What it Offers

pandora®

Pandora offers hundreds of customized stations in more than 40 genres. There are standards such as **Blues, Classical and Comedy** and quite a few unique offerings like **BBQ, Festivals** and **Musica Romantica**.

Setting Up Your Pandora Account

Select the **Pandora** tab from within **Music, Video & Books** to get started. You'll have the option of signing in to an existing Pandora account or registering a new one. A new account requires providing your email, user name, password, gender and zip code.

Once you've signed in or created an account, Pandora is immediately supported by Amazon Dot. For those with an existing account, your stations will appear in **My Stations**, and you can alter or delete them as you wish.

Pandora Basic is free, but short commercials occasionally play between tracks. An upgrade to a Pandora Plus account is available on the Pandora site. A Pandora Plus account is ad free and allows more song skips. Currently, the upgrade is $4.99 per month.

You can use this link to upgrade - **www.pandora.com/upgrade**

Controlling Pandora with Your Voice and App

If you've created a new Pandora account, start by selecting the **+*Create Station*** box at the top of the Pandora page on the Alexa App.

From there, you can use the box to search for an artist, genre or track, or you can see what's already available in the Browse Genre section.

When you select an existing station, it begins to play and also appears in the list entitled **My Stations** on the Pandora Page in the Alexa App.

Select the down arrow to have the option of deleting the station. Pandora allows you to create up to 100 of your own stations to supplement what is already available.

For example, we're fans of a musical genre called reggaeton, and there's not a pre-existing station for it.

However, when we typed reggaeton into the search box, several tracks came up. We selected the track Reggaeton Latino, and a station was created with that tune and similar songs. We spell this out in detail because there are no "how-to's" given on the app. Some tasks, such as creating our own stations, we learned how to do through trial and error.

However, as far as stations are concerned, there are already so many customizable stations within each genre of music, creating your own might not be something you'll want to do anyway.

Any station's playlist can be customized by eliminating unwanted songs from the list. Do this by saying, "**Alexa, thumbs down**", or by selecting the downward pointing thumb on the player at the bottom of the Alexa App.

Note, however, you'll only be able to delete or skip three songs on any list. The fourth attempt will produce a drop-down menu saying, "Our music licenses limit the number of songs you may skip."

In other words, you'll have to listen to the remaining songs all the way through or choose a new station to play.

As ever the player bar at the bottom of the Alexa App offers options to play, pause, go forward and go back. If you click or tap the small icon of the album cover for the song playing, you'll get the full Now Playing page view with a larger icon and a list of the songs as they play.

You'll also have the option here to look at your stations history and the queue of songs that have played on the current station.

Choose the down arrow of any of the songs in the queue to rate it, create a song or artist station, shop for the music at the Amazon Digital Store, Bookmark the song or give the song a rest by selecting "*I'm tired of this track*".

You can use voice commands with Pandora on the Dot, just as you can with other services.

If you want to control content from Pandora with your voice, and we're sure you will, we highly recommend that you choose and add Pandora Stations via the Alexa App first.

We drove ourselves crazy and had a good few laughs trying to get Alexa to play a station if we hadn't already added it via the Alexa App. You can make Pandora the default station service for the Dot but even doing that didn't seem to help much when requesting a station that we hadn't previously saved.

If you don't believe us try asking "**Alexa play GameDay Rap and Hip Hop Radio on Pandora**" and see what you get if you haven't already added that station to your favorite Stations!

It seems that once you have chosen a station and played it once then Alexa is able to recognize what you're saying the next time you ask.

We didn't find making Pandora our default station service particularly helpful but if you want to try go to **Settings > Music & Media**, scroll down and click the **Choose default music services** box. Then choose **Pandora** under **Default station service** and finally click **Done**

iHeartRadio: What it Offers

This network is a digital radio and content streaming service that includes thousands of live radio stations and podcasts from the US. iHeartRadio gives you the option to create personalized **Custom Stations** featuring your favorite artists or genre.

Music is just the beginning. Other categories include **Business & Finance, Comedy, Entertainment, Food, Games & Hobbies, Health, News, Politics, Science, Spirituality** and **Sports**.

Each group offer numerous and wide-ranging options, so it will take you some time to explore what's available and find shows you want to return to often.

We've setup iHeartRadio as our default station service because it offers both curated music playlists, just like Pandora, and real radio stations, just like TuneIn (see below); so two for the price of one.

Setting Up Your iHeartRadio Account

To enjoy iHeartRadio on your Amazon Dot, no account is necessary. Simply use the Alexa App to explore and listen. However, with an account, you can create Custom Stations and share them with others.

Select iHeartRadio on the Alexa App and then click "**Link your account now**" to be taken to a special sign in/sign up page for activating you iHeartRadio with your Dot.

If you don't have an account, set one up using an email address and password. You'll also be asked to provide zip code, gender and agreement with the site's Terms of Service. Another option is to link an existing Facebook or Google Plus account to create an iHeartRadio account.

Controlling iHeartRadio with Your Voice and App

When using the Alexa app click or tap the **iHeartRadio** tab within *Music, Films & Books* and your options appear in the main screen. They include Search for artist or station, Browse for Favorites you've chosen while using the network.

The Browse option gives choices for Live Radio, Perfect For (Kids, Working Out, Driving and more), iHeartRadio Originals and Shows. We had a lot of fun exploring the variety of choices available.

The Originals tab features dozens of unique stations including Golden Era musical numbers, Sippy Cup for pre-schoolers, Classical Genius featuring Mozart and his contemporaries, Workout Beats, All 60s, Road Trip and Chillax.

To use voice commands to play iHeartRadio, you have to know the name of the show, station or radio program you want to listen to.

When requesting a radio station, your best bet to get it working is to refer to it with the exact title of the station as it is shown in the Alexa App.

Some use call letters; others use a name such as The Bear or Kat Country. We had to experiment in order to access some stations.

For example, we requested the station listed as New Country 96.3 KSCS.

When we asked for it by "New Country" the Dot retrieved a top country hits station from Amazon Prime Music.

We then asked for KSCS 96.3, and the radio station was properly accessed.

TuneIn: What it Offers

TuneIn is completely free, and setting up an account isn't necessary. However, having an account allows you to "Follow" your favorite stations and shows and to Share them on Facebook, Twitter and Google Plus and Tumblr.

On TuneIn, you've got the opportunity to access more than 100,000 Internet radio stations including FM, AM, HD, LP and digital. Brands featured include ESPN, NPR, Public Radio International, CBS and C-Span. Your browsing options include much more than Music.

There is Local Radio, Sports, News, Talk and by Location along with popular shows featured in all of those genres. There are more than 4 million podcasts, concerts and interviews available too, a fact that sets TuneIn apart from its competitors.

TuneIn is great but we prefer to use it on our cell phones when we're out and about and prefer iHeartRadio for our Dot.

Setting Up Your TuneIn Account

As mentioned there's no need to set anything up to use TuneIn, you can access and browse everything on offer by following the instructions below.

However, if you do want to take advantage of some further features you can join TuneIn at **www.tunein.com**. Signing up is a brief procedure that includes choosing a user name with an associated email address and a password.

You also have the option of signing in with an existing Facebook or Google+ account, a step that simplifies accessing a new or existing account.

Once you have an account you can create and organize a library of your preferred categories, songs and artists that can be accessed and played on your Amazon Dot using voice or the app.

Similar to Amazon Prime Music, there is a massive amount of content to explore on TuneIn, so enjoy the process of discovery!

Controlling TuneIn with Your Voice and App

If you know the radio station name or station call sign or the name of the podcast, you can use voice commands to get exactly what you want. However, you need to be really precise with the correct name of the station and even then we have had problems.

We had no problems asking "*Alexa play Mix 90's on TuneIn*" but we couldn't for the life of us get "*Alexa play The Bear 98.1 on TuneIn*" to work. In the end we had to use this work around...

If you can't get Alexa to recognize a TuneIn radio station name that you are asking for then instead start playing it manually by clicking or tapping on your chosen station via the TuneIn page in your Alexa App. Then once the station is playing go to the **Now Playing** page.

On the **Now Playing** page, to the right you will see your selected station and a small down arrow (it's called a caret apparently). Click the arrow and the click on *Favorite station* to save it as a favorite. It was only after we had saved The Bear 98.1 as a favorite that we could ask "*Alexa play The Bear 98.1 on TuneIn*" and have her play the station.

Frankly we find it easier to play content from TuneIn using the Alexa App on our mobile phones. From within *Music & Books*, select *TuneIn* and browse for *Local Radio, Trending, Music, Talk, Sports, News, By Location, By Language* and *Podcasts*. Click or tap for dozens of options within each category. When you find what you want, click or tap it, and a player will begin to play the station or show.

SiriusXM: What it Offers

One of the first satellite radio services, SiriusXM now offers streaming Internet radio to compete with iHeartRadio and Tunein. It offers more than 120 channels covering music, news, talk, entertainment and sports plus niche stations like comedy and Latin interests.

Setting Up Your SiriusXM Account

SiriusXM is a Skill, which means that the service can be used with Alexa and your Dot, but isn't supported to the extent Spotify, Tunein and the others are. We have a **chapter on Skills** later, but basically they're third-party services that can be linked with Alexa and used with functionality that might be limited, and that's the case with SiriusXM. A quick click on the Skills section in the app will show you how diverse Alexa Skills are.

All Skills are rated by Alexa users with comments that will help you decide whether to try it or not. The average Skill rating is 3 to 3.5 stars, and SiriusXM is a 3-star skill presently after almost 1,400 reviews. What's playing isn't displayed on your Dot, and SiriusXM sometimes doesn't recognize login information. Some users report the connection is intermittent. If you have SiriusXM already, try the Skill, rate it and let other Dot users know what you think. That's part of the fun of using Skills.

OK, here's how to use SiriusXM with Alexa on your Dot:

- Go to **SiriusXM.com/amazonalexa**
- If you have SiriusXM, select the ***Set Me Up to Stream*** option, and follow the brief instructions on linking your account to Alexa
- If you don't have the service, start a free trial to one of the packages:
    1. 80 Channels called Mostly Music: $10.99/month
    2. 140 Channels called XM Select: $15.99
    3. 150+ Channels including Howard Stern called XM All Access: $19.99/month
- If a new subscriber, open the confirmation email from SiriusXM, and select the link to set up your account
- Log in to SiriusXM when prompted
- Open the SiriusXM tab on the Alexa app
- Enable the Skill, and follow the instructions for linking the account, which might appear in a pop-up window depending on what device you're using the app with.

Controlling SiriusXM with Your Voice and App

Hopefully by now you get the gist of how to operate all these different music services and SiriusXM is no different. Once your account is linked with Alexa you can use your voice to start content playing via your Dot.

Either ask "*Alexa, play (name of station)*" or "*Alexa play channel (station channel number)*"

Full disclosure...we don't use this service and the number of bad reviews for this skill, as well as the added cost, has put us off trying it.

The Now Playing Page Revisited

By this point you should be in a position to play some audio. So go ahead and select a track, either with your voice or manually from any of the services we have discussed in your Alexa App.

Now return to the **Now Playing** page in your Alexa App.

In the main segment of the page you will see a large image indicating the music album, radio station or podcast you are listening to.

Directly underneath you will see a smaller icon/image indicating the source of the audio (Amazon Music, TuneIn etc) and more precise details of what exactly is playing (eg. the title and artist of a track and the playlist/album it is from).

To the right of this information is the volume icon, click or tap on it and a volume bar appears which you can drag to adjust the volume.

And below this are your media player functions which you can use to manually control your media. These are your typical universal controls which include:

- **Shuffle the Queue** of songs lined up to be played by selecting the crossed arrows on the left
- **Loop the Queue**, which means to play it continuously, by selecting the circling arrows on the right
- **Go back one song or skip the song** by using the Back/Forward symbols - solid arrows and line combinations
- **Play the song** by selecting the triangular arrow pointing right or pause the song with the side-by-side vertical lines
- **Adjust the Volume** by selecting the speaker symbol, grabbing the volume indicator and moving it up or down

When a Station or Playlist is playing, you'll have **Thumbs Up** and **Thumbs Down** options in place of the **Shuffle** and **Loop** options.

Amazon will monitor your preferences to customize the Station or Playlist for you.

Of course all these operations can be activated with your voice as described above. Try saying your wake word and then any of these:

- ***Shuffle***
- ***Pause***
- ***Repeat***
- ***Volume 8***
- ***Quieter***
- ***Skip***
- ***Thumbs up*** (for songs on custom Amazon Music stations)

On the Now Playing page you will also see ***Queue*** and ***History***

Queue: This is the list of songs or other media that are lined up to be played in the Playlist, Station or album you've selected. You can select the down arrow next to any of the songs in the list for more options which include:

- Create a Station around the song or artist
- Add the song to your music library
- Shop for a digital version of the song or album on Amazon
- View album in Prime Music (Prime Stations only)
- Rate the song with Thumbs Up or Thumbs Down (custom stations only)
- Select as a Favorite (some podcasts)

History: This list shows all the recent media you've played. Simply tap or select an item from the ***History*** list to enjoy it again. It will then appear as what is **Now Playing.**

6. BOOKS

Now let's take a look at the final section of Music Video & Books page...books. Here you will discover how to listen to Audible audio books and digital ebooks via your Dot.

Audible: What it Offers

audible
an **amazon** company

Listening to books used to be a quirky thing people with long commutes did. Now, the concept is trending and expanding beyond book with a plethora of podcasts (free with a Prime subscription) plus original content created to be read like a radio drama.

We listen to Audible about three times a week, usually to relax in the evening or while getting cozy on a rainy weekend afternoon.

Audible is owned by Amazon, and the company has invested significant dollars in developing content.

Account details:

- $14.95/month after 30-day free trial
- Trial includes one book to keep free, stored in Your Audiobooks library in the app
- 180,000 titles with more added every day
- Many free titles with Amazon, and you can select as many as you like
- Subscription includes one book per month of your choice, and you are given a few days to try the book and exchange it if you don't like it

- 30% discount on additional books
- You have 12 months to return "a set number of books" you don't like, according to Amazon.
- Listen on any Echo device, and control the content with the Player on the Alexa app
- Listen on your smart phone or device using the Audible app
- Use the Whispersync app to listen and keep track of where you are in the book on all your devices

Setting Up Your Audible Account

To set up your Audible account go to **www.audible.com**

- Create an account, and start your free trial
- You will be taken to your Amazon account to sign in
- Select the existing payment method shown by the last 4 numbers or enter a new payment method
- Start your membership, and pick your free book
- Mark your calendar, so you'll remember to cancel within 30 days if you don't want to keep Audible

Connecting Audible and Alexa:

- Choose *Audible* from the *Books* section in the Alexa app
- Choose *Link Account*, and follow the brief instructions
- Your Audible book or books will appear in the *Your Audiobooks* list

Controlling Audible with Your Dot and App

- Manually choose the book you want from Your Audiobooks library or say, "*Alexa, read The Magic of Thinking Big*"
- For books previously begun, say "*Alexa, resume The Magic of Thinking Big*," and it will pick up where you left off

- The book will appear on your Dot, on the Now Playing page and on the small player at the bottom of any page when viewing the app online
- Manually select a chapter on the Player or verbally request the chapter you want
- Use the Player to go back or forward 30 seconds, pause or play and change the volume
- Use verbal requests like "*Alexa, go back two minutes*" or "*Alexa, go to the next chapter*"
- Request Alexa to "*stop reading in 10 minutes*" or "*set a sleep timer for 10 minutes*"

Kindle Books: What it Offers

amazon kindle

Kindle has dominated the eBook market for a decade, and now Alexa can read many of your favorites with hands-free ease.

- You can buy individual titles without a subscription by choosing the Kindle version when purchasing a book
- Your second option is a Kindle Unlimited subscription for $9.99/month with unlimited access to one million books and magazines, though few are new or recently released books
- A six-month subscription is available for $59.94
- Amazon Prime members get one early release book called a Kindle First plus one additional book free each month
- Read books in the Library of your Amazon Household members (See **Chapter 4** for details or go to the **Accounts & Lists** tab on **Amazon.com** and select *Amazon Household* in the *Shopping Programs and Rentals* section)

- You have the option to borrow and lend books with other Kindle users and rent textbooks too
- Personally we're not fans of Kindle Unlimited, we tend to buy books by well-known established authors and rarely find their books included in the subscription. Furthermore, while we like this audio feature on Alexa, the books are read in Alexa's computerized voice which is a poor substitute for a professionally recorded audio book.

Setting Up Kindle Books Audio

- No account is needed to buy individual books
- If you're a Prime member, If you're a Prime member, you already have a Kindle First plus one additional book free each month, and it can be accessed from the **Accounts & Lists** drop-down menu on **Amazon.com**
- If you already have a Kindle account, your Kindle books will appear in your Kindle Library on the Alexa app
- For a 30-day trial to Kindle Unlimited, go to **Amazon.com/kindleunlimited** or type Kindle Unlimited in the Amazon search box
- Choose Start your 30-day Free Trial
- Sign into your Amazon account
- Choose an existing payment method or enter a new one
- Start your membership, and enjoy browsing for titles that interest you
- Books you buy, borrow or rent will show up in the Alexa App
- Mark your calendar, so you'll remember to cancel within 30 days, if you wish

Controlling Kindle with Your Dot and App

Since Kindle is an Amazon service, it's already connected to Alexa. Choose **Kindle Books** in the Alexa app, and your available books should be ready to enjoy. Books are read in Alexa's voice.

- Manually choose a book from the list called **Books Alexa Can Read**, and it will start

- Control it on the **Now Playing page** of the Alexa app including selecting the chapter you want from the **Queue**, changing the volume, pausing it or going back or forward 30 seconds

- Optionally use your voice to request a title, a chapter of the book currently being read, or say, "***Read a Kindle book***," and Alexa will ask for the title you want to hear

- Once the book is being read, you can say things like, "***Alexa, go back three minutes***", "***Stop reading in 10 minutes***" or "***Set a 10-minute sleep timer***" to stop the reading when you choose

- If you've started a book previously, Alexa will start where you left off.

Audiobook Immersion Reading: We wonder whether this concept will catch on. Immersion reading is listening to a book being read while viewing the text.

The text is highlighted as it is read. Amazon suggests that immersion reading creates a deeper connection with the book. We've tried it, but found it oddly difficult to read and listen at the same time, even to the same text. If you're curious and want to learn more or try immersion reading, visit **audible.com/mt/immersion**.

7. LISTS

Amazon first envisioned Echo devices as virtual personal assistants, and this remains a significant part of their core identity. The Alexa app and Dot work well together in helping us assemble, view and manage our Shopping List and To-do List on a daily basis, and we're sure they'll do the same for you.

The Dot personal assistant does a pretty good job populating the list via voice commands, though it might get a word wrong here or there. We asked, "**Alexa, add 'find CDs' to my to-do list**", and what appeared on the list was, "**Find c.d.s**". Then, a note to get "**screen grabs**" was listed as "**screen grahams**." Misunderstandings happen most often with uncommon words.

There have been occasions where we've had to cast our minds back and remember what it was we asked Alexa to add because what's on the list doesn't make sense. It's inspired a blend of comedy, frustration and "ah-ha!" moments when we figure it out.

Typically, a request such as, "**Alexa, add mow the grass to the to-do list**" is clearly understood. Amazon suggests as an alternate saying something like, "**Alexa, I need to organize my tools**". We've found that the Dot doesn't comprehend this voice command very well. We stick to the "**Add (something) to my to-do list**" commands for best results.

Select **Lists** in the Menu where you will see your **Shopping List**. The heading next to it is your **To-do List**. Manually manage the lists by clicking/tapping back and forth between them.

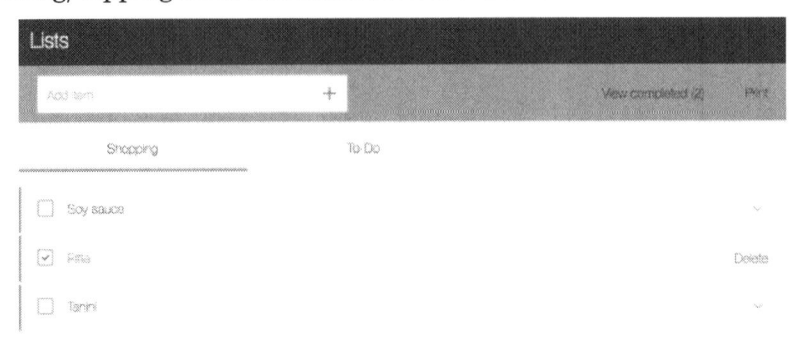

Shopping Lists

Creating a shopping list has never been so easy. Add items as the need arises, and you'll have a list you can print or take with you to the store on your mobile device.

We haven't shopped from a paper list since buying our first Amazon Echo a few years ago.

Our list is right there on our phone, and using the app, we can also compare prices at Amazon for major items we're looking at in a store.

Create and add to your list in two ways:

- **On the app:** Type what you need into the **Add Item** box, and select the **+ *sign*** or hit ***Enter***, and it will be added.
- **Using voice:** Tell Alexa to add what you want to the Shopping List, and she'll tell you it was added. Alexa will add doubles of things, if you ask for it twice. That's a minor issue easily remedied. Removing items is discussed below.

When using Alexa on your computer, choose the ***Print*** option, and your computer's **Print Manager** will open.

Review your Shopping List: Check it any time on your computer or device using the app. Say, "***Alexa, shopping list***," and Alexa will read the list to you.

Manage your Shopping List in these ways:

- **On the app:** Check off items by selecting the box. Select ***Delete*** to remove them immediately, which we do when we mistakenly add something a second time, or they will be transferred to the **Completed list** when you navigate away from the Shopping List. Select the down arrow for these options:
  1. ***Search Amazon*** for tahini (more about online shopping later)
  2. ***Search Bing*** for tahini

3. ***Move item to To-do List*** (If you say, "***Add clean the windows to the list***," Alexa might add it to your Shopping List. Be sure to specify the To-do list for that sort of thing.)

4. ***Delete Item*** to remove something from the list

The list on the app shows the Active list, that is, things not yet checked off. You can view checked-off items by selecting View Completed at the top of the page. There, you can also delete checked off items individually or all at once.

- **On Your Dot:** Say, "***Alexa read me the shopping list***" or similar request, and Alexa will read it to you.

Slightly annoyingly, if you want to remove an item you have to do it via the Alexa app, you can't do it with your voice.

A Card is created on the app **Home** page when you add an item to the Shopping List. Options on the Card include **View Shopping List** (and you'll be taken to it) and the same shopping options available on the List such as searching Amazon or Bing for the item.

How we use the Shopping List: The great thing about these options is that we can compile a single list using multiple forms of input.

If we're going through the refrigerator or the pantry, it's easiest to verbally list items. However, we suggest listing them one at a time.

For example, it's easier to say, "***Alexa, add romaine lettuce, red peppers and kale to the Shopping List***," but then they all appear as one item, not three. When shopping, we prefer each item listed separately to avoid overlooking something.

When we're away from the Dot, adding items to the list via our phone or laptop is very easy.

Frankly, if you're a fast typist, you might be able to add a bunch of items to your list manually much faster than you can using voice to mention an item, wait for Alexa to respond, mention an item, wait... and so on.

AMAZON ECHO DOT USER GUIDE

To-do Lists

It's just as easy to create and manage a To-do list. The process is much the same.

Create a To-do list: You can start a list and add to it in two ways:

- **On the app:** When you select the *Lists* tab on the app, your **Shopping list** will appear. Choose the *To-do* tab next to the Shopping heading, type the task into the *Add Item* box, and add it by selecting the **+** *sign* or pressing the *Enter* key.

- **Using voice:** Say, "*Alexa, add reserve meeting room to my to-do list*." Alexa will let you know the task has been added.

Review your To-do List: Select the *To-do* tab on your **Lists** page or ask Alexa to "read" your list to hear it. If you simply say, "*Alexa, to-do list*," Alexa will ask, "What shall I add to your To-do list."

Manage your To-do List in two convenient ways:

- **On the App:** Select back and forth between the Active list of uncompleted tasks and the Completed list of checked tasks. On the Active list:

  1. Check off items by selecting the box, and a check-mark will appear

  2. Delete checked items immediately, if you want to remove it

  3. **View more options:** Select the task itself, not the check box, to move item to Shopping List or Delete Item

When you navigate away from the Active list, completed items that have not been deleted will show on the Completed list. Select "*Delete all*" from the completed list or delete them one at a time.

You can also deselect the task (uncheck the box), and the task will return to the Active list.

- **With Voice:** Say, "Alexa, read my To-do list" or "Alexa, what's on my To-do list," and your Active list will be read:

If you say, "*Alexa, remove wash the car from my To-do list*," Alexa will give you verbal instructions on how to do it manually on the app.

Cards Created for List Items

A card is created on the app **Home** page when you add an item to the To-Do List. A box on the Card can be checked, and the item/task will be checked on the appropriate list in the app. The Card also offers the options we mentioned with each card such as "Move item" to the other list.

How we use the To-do List: We compile the list much like our shopping list - using all options. If one of us knows a handful of things that need doing, manually adding them on the app via the computer is easiest. Things that come to mind one at a time are more easily added by voice when at home and via the phone and the app when away.

8. REMINDERS & ALARMS (AND TIMERS)

We think you'll find these Alexa tools very useful, as we do – on nearly a daily basis.

When you select the ***Reminders & Alarms*** tab on the Alexa app, it opens to show a page with three options across the top: **Reminders, Alarms** and **Timers**.

The Reminders page is shown. Select either **Alarms** or **Timers** to use those functions. Let's review them individually.

Reminders

How often do you forget something, even a minor task around the house, that is time sensitive?

That's the key for us: We use Reminders when time is a factor. Otherwise, we add tasks to our To-do list. Sometimes we use both tools to be sure we do what we need to do and on time too!

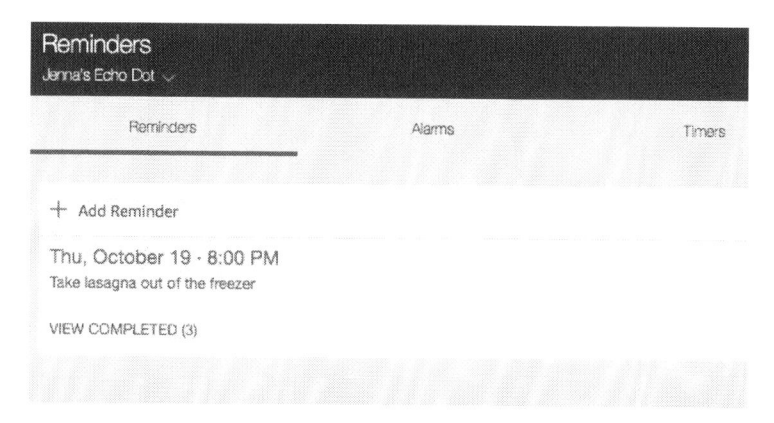

```
Reminders
Jenna's Echo Dot

    Reminders              Alarms              Timers

+ Add Reminder

Thu, October 19 · 8:00 PM
Take lasagna out of the freezer

VIEW COMPLETED (3)
```

Create Reminders: As with lists, you can set Reminders manually and using voice.

On the App: Typing in the Reminder ensures that the details are correct:

- Click or tap the blue + ***Add Reminder***, and the Reminder form will appear.
- In the "***Remind me to…***" box, type in the specific reminder you want to hear, up to 128 characters.
- Select the ***Date*** tab, and today's date will appear in xx/yy/zzzz form. If the Reminder is for another day, select ***Date*** to change it. The month will be shaded blue. If that is correct, use the tab key or forward arrow key to move to day or year. Once you move on from the date, it will appear with the month spelled out.
- Select ***Time***, and the current time will appear. Change it as you changed the date.
- Your device's name is shown. If you have more than one device on your Amazon account, select the down arrow to choose the device you want to give the Reminder.
- ***Cancel*** or ***Save*** the Reminder.

Using Voice: This is a much faster method of creating a Reminder. You can create Reminders using a specific time or asking for the Reminder to be made in a set amount of time:

- Say, "***Alexa, Remind me to take lasagna out of the freezer at 8pm***", or "***Remind me to leave for the show in 45 minutes***."
- If you don't give a time, Alexa will ask you for one. Say, "***8pm***" or "***in 45 minutes***." When giving specific times and you forget the AM or PM, Alexa will ask, "Is that 8 this morning or evening?"
- If you give a specific time but no date, the Reminder will be set for today. To give a date, you can say things like "***today***," or "***tomorrow***." If you say, "***Friday***," the Reminder will be set for the next Friday, and if you don't give a year, it will be set for the next time that date occurs. Of course, you can also give an exact time and date such as "***4pm on December 15***."
- Alexa will affirm the reminder with, "OK, I'll remind you at 8pm" or "in 45 minutes."

Review and Manage Reminders:

- Manually select the **Reminders & Alarms** list to view your list of **Reminders**, or say, "*Alexa, read me my reminders*" or "*What are my reminders?*"
- Alexa will read the list of Reminders
- You must manually edit a reminder; voice can't be used.
- Verbally cancel a Reminder by saying, "***Alexa, cancel the reminder for 8pm,***" and Alexa will confirm the cancellation or ask a question to clarify.

Here's a tip that applies anytime you converse with Alexa: When Alexa asks you a question and the light bar is blue/green, you don't have to use your wake word in your answer. The light bar indicates the device is listening.

- Choose ***View Completed*** at the bottom of the list to see past Reminders
- To Edit or Mark as Completed, select a Reminder from the list. Those two options will appear, and you can select your choice. When you select ***Edit***, the form will appear. Select what you want to change, make your edit, and select ***Save Changes.***
- To Delete the Reminder, select ***Edit*** and the ***Delete Reminder*** option

Getting Reminders: At the Reminder time, Alexa will say "It's 8pm. Take lasagna out of the freezer." A chime will sound. This happens twice unless you ask Alexa to cancel the timer.

Completed Reminders: On the Reminders page you can click ***VIEW COMPLETED***. Reviewing them might help you remember whether you followed through on the action. Of course, the list only means that the Reminder was given, not that you took the lasagna out of the freezer! Adding the task to your To-do list as well and checking it off gives more assurance that it was done.

For some reason Completed Reminders can't be deleted they just remain within the Completed page. To return to Active reminders from here, you must select the ***Reminders & Alarms*** tab.

Alarms

The Alexa app has a few quirks, and one of them is that Alarms must be set verbally; there's no manual way to do it, probably because it's so easy using voice.

Creating an Alarm using Voice:

- Say, "*Alexa, set an alarm for 6:45am.*" If you forget to specify any of the details, Alexa will ask.
- Alexa will affirm the alarm details.
- You can also say, "*Alexa, set an alarm for 6:45am every morning*" or "*Set an alarm for 6:45am weekdays*" or "*weekends*" or even "*Set an alarm for 6.45am every Monday*"
- You can set multiple alarms for each day or different alarms for each day of the week.
- One-time alarms must be set for a time within the next 24 hours.

Managing Alarms: Alarms can be managed manually and with voice:

Using Voice:

- Say, "*Alexa, cancel the alarm for 6:45am*" whether it is a one-day or multi-day alarm. It will be turned off, and Alexa will affirm it. You cannot turn off a multi-day alarm for just one day. If you have a 6:45am weekday alarm, and you say, "*Alexa, cancel the 6:45am alarm for tomorrow,*" Alexa will say, "you don't have an alarm set for tomorrow morning." Saying, "*cancel the 6:45am alarm*" will turn it off for all days.

On the app: To see your Alarms, select the **Reminders & Alarms** tab and then **Alarms** from the three options at the top.

- Any alarm you've set, even if it has gone off or been cancelled, will appear in the list. You can toggle it on or off.
- To edit an alarm, select it from the list. Your editing options include time and frequency of alarm including options like

Every Day, Weekends, Every Monday, etc. You can Delete the alarm there or select **Save Changes**, if it has been edited. If you don't make changes, select **Cancel** or simply click away.

- The current Alarm Sound will show. Select that box to see options for **Celebrity Voices** and for **Custom Sounds**. When you select one, a sample sound will be played on your computer or phone, but not on the Dot. Celeb voices are fun, and Custom Sounds range from interesting (Rainier) to quirky (Adrift) to ethereal (Squared Waves) to very annoying (Countertop). We keep it fresh by mixing celeb voices in with standard sounds too.

Manage alarm volume and default sound: This option is shown above the list of alarms. Select it to:

- Review the alarm volume or drag it left (quieter) and right (louder) to adjust volume.
- Review and change the alarm sound.

The audio for notifications is discussed later. The Request Sounds option is quite straightforward.

If you select **Start of Request**, a single tone will sound when you say the wake word to let you know Alexa is listening.

Turning on **End of Request** will cause Alexa to sound a tone when the request is complete.

Timers

Timers are easy to set and come in quite handy for obvious uses like baking, but we also use them to be reminded it is time to leave for an appointment, a game on TV is about to start and many other one-time tasks. They're much like Reminders, but even easier to set.

Timers is the third option at the top of the **Reminders & Alarms** tab. They must be set with voice.

Setting Timers:

Say, "***Alexa, set a timer for 20 minutes***" or any length of time from a second to 24 hours.

- Alexa will say, "20 minutes starting now".
- During that initial display, the Timer can be swiped left to cancel.
- More than one Timer can be set, and they can be named, if that helps you keep them straight! Say, "***Alexa, set an oven timer for 90 minutes***" or "***Set a rice timer for 60 minutes.***"
  This is invaluable when you're cooking a multi-course meal or when using timers for cooking and for other tasks too.

Here's a tip: Using Reminders concurrently with Timers can be very helpful too. For example, when we know the dish in the oven needs 90 minutes, we set a timer. Then, we'll set Reminders such as, "***Alexa, remind me to start the rice in 30 minutes***" and "***Remind me to put the bread in to warm in an hour.***"

Viewing and Managing Timers: Use voice and app to manage timers.

View timers on the app where they will be listed with a countdown in progress.

- **Using voice:** Say, "***Alexa, cancel the 20-minute timer***," and Alexa will say, "***20-minute timer cancelled.***"
- **Using the app:** Select a Timer from the list, and select the **Cancel** option. You can also select **Manage Timer Volume** to slide it left for quieter and right for louder.

Canceled and completed Timers do not remain in the list as canceled and completed Alarms do.

AMAZON ECHO DOT USER GUIDE

9. ROUTINES

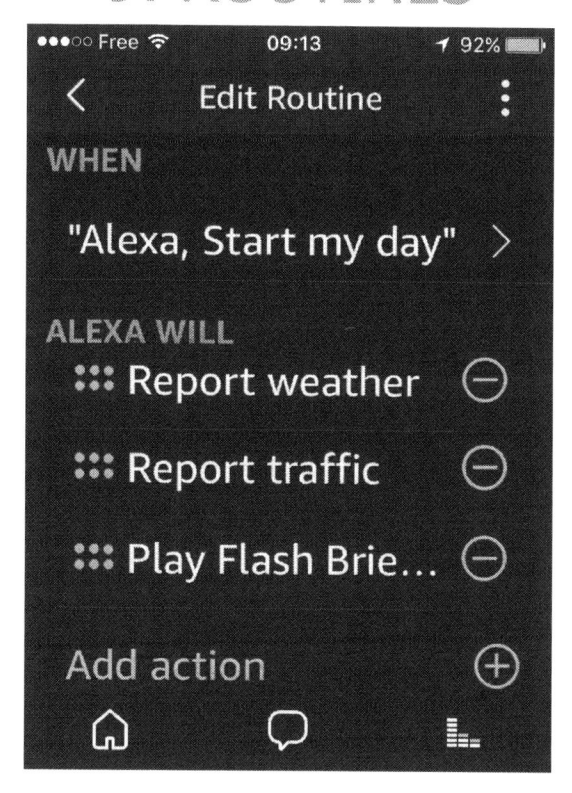

A few words to Alexa can now put a series of customizable functions on auto-pilot, so to speak. The feature is called Routines, and here's how it works: You create a Routine once, select a request phrase or a time of day to trigger it, and the routine's functions occur without having to make separate requests for each one. Currently, though subject to change as all things with Alexa seem to be, Routines only works with:

- Smart Home automation devices
- Flash Briefing (News, sports, entertainment news, etc.)
- Weather information
- Traffic/Commute information

Each of these Alexa functions are discussed in other chapters, so reviewing them first will assist you in making the most of the possibilities available with Routines.

Still, its possibilities are limited. It would be helpful if a reading of your day's calendar schedule, playing the Tonight Show monologue from Skills (or starting any of the Skills you use), tuning to a specific radio station on Tunein or Playlist on Amazon Music were possible too, and we suspect features like that will be added to Routines.

Our first foray into Routines was an evening routine that goes like this:

We say, "**Good evening,**" and...

- Alexa turns on automated interior and exterior lighting
- Echo Dot plays our Flash Briefing to catch us up on the news of the day

We have a morning Routine too. At 7am, it starts our coffee pot, adjusts our smart thermostat a few degrees and plays our Flash Briefing, which starts with several sports briefings, and we often stop it after them because we peruse the news online while we have breakfast.

A nighttime Routine that adjusted the thermostat and lights, locked doors and turned off the TV would be useful too, though we don't have enough home automation equipment to make it worthwhile yet.

Here's another "as of now" oddity we expect will change: There isn't a Routines tab on the Alexa app for PC; It can only be accessed on mobile apps.

OK, let's talk specifics on setting up Routines. We suggest that you go through this step-by-step with us because, frankly, the instructions make it sound more complicated than it is. You'll see how easy it is, if you follow along.

The first step is to choose **Routines** from the main Menu on your mobile device. From there:

Tap the + **sign** at the top near Routines, and two options will show: **When this Happens** and **Add Action**

The **When this Happens** is the trigger, and when you tap its **+ sign**, you'll be given two options: ***When you Say Something*** and ***At Scheduled Time***

Selecting ***When you Say Something*** will open a keypad where you can type in the word or phrase to invoke the routine

Select ***At Scheduled Time***, and you'll see: **At Time** and **Repeat options**

We tapped the ***At Scheduled Time*** option to select 7am on the clock dial that appears before selecting **OK** and **Done**

Then we tapped the ***Repeat*** option to choose ***Weekdays*** from a list that also offered Weekends and individual days of the week

In true Amazon fashion, you can choose Weekdays, Weekends or ONE day of the week, but not two or more individual days of the week. Complete the trigger side of the equation by selecting ***Done***.

Now you're finished with the trigger part - a word/phrase or time. The next step is to decide what happens. Select ***Add Action***, and four options will appear: **News, Traffic, Weather** and **Smart Home**

Choose the ones you want, one at a time, and follow the easy prompts to customize them before selecting ***Add***

Once you've set up the Routine, select ***Create***, and it will be scheduled

Your Routines show in a list on the main **Routines** menu page. Select a routine to modify any of its options, delete it or toggle it Off (and later, back On, if desired).

As mentioned, setting up smart home skills, and traffic, news and weather briefings are discussed later in this guide so become familiar with these options before setting up any routines.

10. ALEXA SKILLS

With Skills, the potential to personalize your use of Alexa and the Dot is tremendous.

An Alexa Skill, in simplest terms, is a third-party function that can be used with Alexa and the Dot - Find a recipe, order an Uber, ask quiz game questions, listen to the Tonight Show monologue, order pizza, find your phone, lead you through a short workout, turn on outdoor lighting... the spectrum is very diverse. Most Skills are easy to set up and use; there's a bit of a learning curve for a few of the more technical ones.

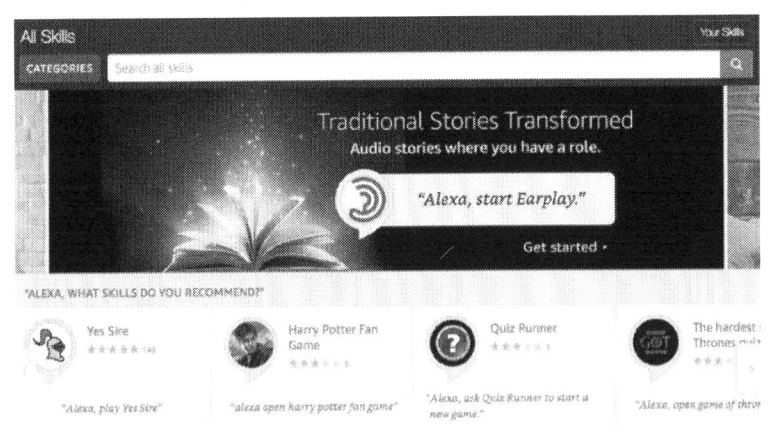

This guide will help you understand the basic concept of Skills, give tips for putting Alexa Skills to work for you and suggest a few to try.

Enabling Alexa Skills

Let's go over the basic approach at the start to show that it's fairly easy:

- Find a Skill you want to use
- Enable the Skill by selecting the blue **Enable Skill** box (which will switch to Disable Skill when enabled)
- For some Skills, you will be asked to follow prompts to link an account or instructions for setting up the Skill for use, such as

entering your phone number for a Find My Phone skill (Most of these types of Skills say **Account Linking Required** under the **Enable** button)

- Grant permissions (very few Skills require this) such as accessing your Echo device address to make the Skill more functional
- Once the Skill is active, ask Alexa to put the Skill to use, such as, "**Alexa use Find my Phone**" (There are "Say…" tips for every Skill)
- If the Skill is more than a one-time event like playing Jeopardy or listening to a radio station, control the Skill with voice requests such as **Pause, Resume** and **Stop**
- Disable a Skill by selecting it from the **Your Skills** list and choosing the **Disable** option
- (Optional) Rate the Skill and add your Review to assist others in picking useful Skills and avoiding duds

The list of Skills you have enabled is viewed by selecting the **Your Skills** tab at the top of the main **Skills** page on the app.

Deciding which Skill to try can be something of a challenge - harder than using most of them! There are more than 25,000 Skills divided into 21 categories. Many new Skills are added daily. Don't expect to browse them all - It's like walking into a bookstore and trying to read the front cover of every book on the shelves. That's unrealistic. Here's how we discover Skills:

- **Use the Search Box at the top:** This is always our first choice. It allows us to find exactly what we want or a Skill related to our interests. For example, one of us wants to brush up on the Spanish language skills learned in college, so the term "**Spanish lessons**" was typed into the box. Three Skills were displayed as results. We enabled two of them, tried them out, and disabled one.
- **Select the Categories Tab at the top:** Here, you can browse categories that interest you. We sometimes look at the New Arrivals category to try a few. When you select a Category, here's what you'll see:

1. The option to search within the Category: This saves a lot of time if you're looking for something specific.

2. The number of Skills in the Category

3. **Sort by:** The default Sort By method of listing Skills in each category is by Featured Skills. Select Sort By to change the listing to Relevance, Average Customer Rating or Release.

4. **Recommended for You:** We're never sure why the Skills in this list are recommended for you. Sometimes there seems to be a connection with other Skills we have Enabled. The arrow to the right of the list allows you to scroll and browse horizontally.

5. **Top Enabled Skills:** These are popular Skills with high ratings, usually 4.5 to 5 Stars. We rely heavily on the recommendations of others.

6. **A vertical list of the Skills in the Category:** These seem to be recently added Skills and Skills Amazon wants to feature.

- **Scan the Rows:** Each horizontal row has a different title, and rows can be browsed using the scroll arrow at right. Some of the many rows currently showing are titled:

    1. **New Skills:** Newest "big name" Skills, currently Skills from Rosetta Stone, Nissan, Lexus

    2. **Top Skills:** Skills that are highly rated and popular

    3. **Recommended Skills:** Skills Amazon thinks fit you

    4. **My Skills:** Your enabled Skills

    5. **Game Skills:** Quizzes and similar games

    6. **Morning Skills:** Starbucks Reorder, Daily Affirmation, Morning Music and similar

    7. **Food and Drink Skills:** Recipes, Restaurants, Cooking Tips, etc.

Skills on Amazon.com: Skills can be searched, browsed and enabled at Amazon.com too. Hover over Departments, then hover down to Echo & Alexa. When you do, a topic list called **Content & Resources** will appear. Select *Alexa Skills* from the list. Or simply go here - **www.lyntons.com/AmSkills**

Tips for Picking Skills to Try and Managing Enabled Skills

These tips about deciding which Alexa Skills to try first should prove useful to you, as they have for us:

- **Read About This Skill and Skill Details:** When you select a Skill to consider it, read both these sections to learn what it does, how it does it and how the Skill is used. You'll have a clear idea whether the Skill is a good fit for you.

- **Go by Ratings and Reviews:** The average rating on Skills, as you can see, is about 3 Stars. We don't bother even trying Skills with fewer than 3 Stars. Many don't work well are simply aren't very useful. We much prefer 4-Star and above Skills. It's also helpful to open the Skill to read the reviews. Knowing why people like or dislike the Skill will help you decide if it will suit your purposes or not. This tip is very helpful when there are multiple Skills for many purposes like finding your phone (11 and growing) and finding a local restaurant (5).

- **Most Skills waste time:** It's easy to lose an hour or more browsing and trying Skills - and then wonder where the time went. Most Skills are not particularly useful. We're at the point where we only go looking for Skills if we have a specific purpose in mind like relearning Spanish. Sure, we have a few "fun" Skills enabled, but we find we use them less and less.

- **Disable Skills you don't like:** As soon as you realize a Skill doesn't work, wastes your time or in any other way isn't a good fit for you, disable it.

- **Disable Skills you don't use:** We go through our Skills once a month or so to weed out Skills we don't use. This unclutters our list and helps us find the Skill we do want at any given moment.

Develop a Skill to Share

Alexa Skills are developed by Fortune 500 companies and the guy living next door who does it for a hobby. Anyone can develop a Skill and release it to be used, rated and reviewed.

If that intrigues you, see the Amazon Developers Site that's all about creating Skills. Check out **developer.amazon.com** and select *Alexa*.

There, choose *Get Started with Alexa* to access what Amazon calls the Amazon Skills Kit, or ASK.

You have the option to use a Skills template or build one all your own. Developers are rewarded with perks from Amazon ranging from developer shirts to Echo devices.

11. SMART HOME

Smart home enthusiasts enjoy using Alexa because it allows them to control their smart home equipment with hands-free convenience. Wink, WeMo, Nest, ecobee, Phillips, SmartThings - Alexa integrates with most smart home equipment in five easy steps using Skills, the subject of the **last chapter**.

Setting Up Your Smart Home Devices

Here are the steps, all explained in detail below:

1. Verify that your smart home device works with Amazon Alexa by finding an Alexa Skill for it

2. Prepare the device for linking to Alexa

3. Enable the Alexa Skill for the device

4. Ask Alexa to discover the device, so it can be controlled using Alexa

Once you've setup your devices, and connected to Alexa, we will discuss how to get the most out of your smart home products by grouping them into Groups and Scenes.

Before we get into the details of these steps, it's always a good idea to put safety first. Amazon recommends:

- Use the smart home equipment only as recommended by the manufacturer, and follow its instructions

- Check to see that requests such as locking doors, closing the garage door and turning on security lighting have been carried out

- Once a device is connected to the Alexa app, it can be controlled by anyone who uses your Echo Show, so be sure to instruct everyone in the household about the smart device's proper use

- Since home security is critical, consider turning off Echo device microphones to prevent unintended changes to locks, lighting and other security features

Let's work through these steps systematically giving step-by-step guidance where needed.

1. **Verify Device Compatibility with Alexa**

If it's compatible with Alexa, there will be a Skill for it:

- Select the **Skills** page within the Alexa app
- Type in the device name, such as WeMo, ISY or Samsung SmartThings
- Hit *Enter*
- Locate the device's Skill in the handful of results you get

Another option is to visit the Amazon Smart Home webpage and search that department for your smart home device. Devices compatible with Alexa will have a small "**Works with Amazon Alexa**" white banner on their product page. It's also worth visiting this department just to get a sense of what's available in the Smart Home marketplace right now. Visit - **www.lyntons.com/SmartHome**

2. **Prepare the Device for Linking to Alexa**

With the Skill located to verify compatibility, now it's time to get your smart home equipment ready. This is done independently and away from your Echo Show or Alexa app. Basically, you need to get any smart home device up and running correctly first before linking it with Alexa.

- Download the smart device app (available on the manufacturer's site) to a mobile device
- Use the device app to set up the smart home device on the same Wi-Fi network that Alexa is using (viewable on the Alexa app by going to *Settings > Your Echo > under Wireless*)
- Once the smart home device is online, download and install software updates if prompted

Note to Phillips Hue Bridge users: At this writing, the Hue Bridge works with Echo, Dot and Tap, but not Show. That might have changed by the time you read this, since the V2 (second generation Bridge) has been released. Check for a Skill or try pressing the Bridge button to see if it can be found by your Echo Show.

3. **Enable the Alexa Skill for the Device**

Now that your smart device is ready to go, enable its Skill, which you located in step 1 when verifying compatibility:

- Select the **Enable** box, and it will change to **Disable**, indicating the Skill has been enabled
- Read the information on the device's Skill page to discover the easiest and most effective way to use it with Echo Show including suggested voice commands
- When you later want to find the Skill, it will be in the Your Skills list which you get to from the Skills page of the Alexa app.

4. **Ask Alexa to Discover the Device**

There are two approaches:

- Say, "*Alexa discover devices*," and Alexa will search for them and report to you what was found
- Go to the **Smart Home** tab on the Alexa app, select **Devices**, and select the **Discover** box there
- **Note:** The Enabled Skills for your smart home device should also be listed on the Smart Home page on the app

Getting the Most Out of your Smart Home Devices

You should now have all your smart home devices connected to Alexa and be able to see them all within the Smart Home page of your Alexa app. So now let's look at bundling them together by placing them in

Groups and/or managing Scenes to get the most functionality from your automated household

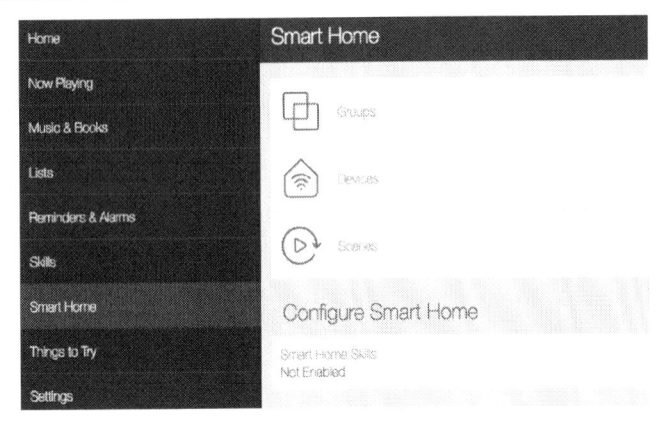

Organize Smart Home Groups:

Putting devices into a Group allows you to turn multiple devices on or off with a single voice command.

Here's how to create and manage a group:

- Select the **Smart Home** tab from the Menu
- Select **Groups** and then **Add Group.** You will be given two options, you should select **Smart Home Group**. If you have multiple Echo devices then you can learn more about the other option here (**Multi-Room Music Group**) in our **Help & Feedback chapter**.
- Create a name for the group that is at least two syllables, distinct from other group names and spelled properly. Alternatively, select one of the common group names that Alexa offers.
- Pick which devices you want in the group
- Select **Save**
- Try controlling the group with a single command such as "*Alexa, turn on Living Room*," or "*Alexa, lock doors*."
- Edit a Group using the Add/Remove devices option in the Groups section of the Smart Home page, or select Delete Group to eliminate it

Create and Manage Scenes:

With Groups you are simply putting different devices together so you can switch them all on and off at once. Scenes are a little more sophisticated and harder to quantify in this guide because they are specific to the smart home devices that you own.

So a Scene is an action/feature that you have set up separately within the manufacturer's app of your smart home devices. When you connect that device to Alexa any Scene that you previously setup should now appear in the Alexa app when you navigate to **Smart Home > Scenes**

To manage all the available Scenes select the Scenes tab on the Smart Home page in the Alexa app

- Follow the prompts to customize the Scene produced by your smart devices
- Select the Scene to edit or delete it

See Skill page of each device for suggestions about controlling it using voice. Most are straightforward and predictable. For example, say: "*Alexa...*"

- T*urn on or off a smart home device, group or scene"*
- *Brighten or dim a single light or group"*
- *Set the light to pink/soft white"*
- *Set the thermostat to 70 degrees"*
- *Lock the back door or lock the doors"*

Troubleshooting Smart Home Setup

Try these Alexa smart home tips in this order, starting with the most likely problems:

- Double-check compatibility of the smart home device with Alexa
- Make sure the device's Skill is enabled
- Download the latest device software update

- Restart Wi-Fi, your Echo Show devices and the smart home device(s)
- Disable and Enable the Skill
- Unlink the smart device from Alexa using the "*Forget*" option for the device
- Follow the setup steps above to relink Alexa and the device

Troubleshooting Wi-Fi Issues with Smart Home Equipment

The most common problem is that the Echo Show and the smart home devices are not on the same Wi-Fi network. If you want to change the network the Show is on:

- Go to **Echo Show Settings**, by sliding down on the screen, starting above the display, and choosing **Settings** from the Menu
- Check the Wi-Fi network that the Echo Show is on - it's listed under the **Wi-Fi** tab
- If the Echo Show is not on the network the smart devices are on, tap the **Wi-Fi** option, tap the correct network and type in its password before tapping **Done**
- Optionally, and if this is something in your technological skill set, tap the **Add a Network** option in the **Wi-Fi Settings**, and add a network for Echo Show and the smart home devices (for preventing interference from other devices on the network)

Try to Discover Devices Again: After making any necessary changes outlined above, ask Alexa to "*discover devices*" again. If the reply is "I didn't find any devices" or the device in question wasn't discovered, contact the smart home device's manufacturer. After all, it is up to the device manufacturer to ensure its equipment is compatible with Alexa.

Top Smart Home Alexa Skills

There are about 700 Skills in the Smart Home category, and like Skills in all categories, some are more useful than others. Here are some Skills that work as advertised and offer genuine benefit, in our opinion:

Ecovacs Deebot (3.5 Stars): Allows you to control most Deebot vacuum models using Alexa. Say, "*Alexa ask Deebot to start cleaning*" or "*Ask Deebot to start charging.*"

Kenmore Smart (3.5 Stars): Allows you to monitor and control some Kenmore Smart and Elite home appliances. Say, "*Alexa ask Kenmore Smart what is my freezer temperature*" or "*Ask Kenmore Smart to start the dryer.*"

ADT Smart Security (4 Stars): Allows you to monitor and control your ADT home security system. Say, "*Alexa ask Smart Security to run 'daytime'*" or "*Alexa, ask Smart Security to run 'dinner party.'*"

To find Skills for your smart home equipment, search for your equipment brand in the Smart Home Skills category. Give the Skills a try. If they're useful, you've added convenience and functionality to your life. If not, disable them and try something else. That's how we do it, and it allows us to find Skills we use rather than those that clutter our Skills list waiting to be disabled.

12. THINGS TO TRY

You won't use Alexa and Dot long before you realize Amazon is constantly suggesting something for you to try.

That's what this chapter is about. Most of the topics in the **Things to Try** section of the app are clear and straightforward.

Our goal in this chapter is to uncover things that aren't as clear and give you inside tips from experienced Alexa users that should assist you in getting the most from your Dot.

The Things to Try section is just a list. When you select any of the topics, you're taken to another part of the app, usually **Help & Feedback** or **Settings**.

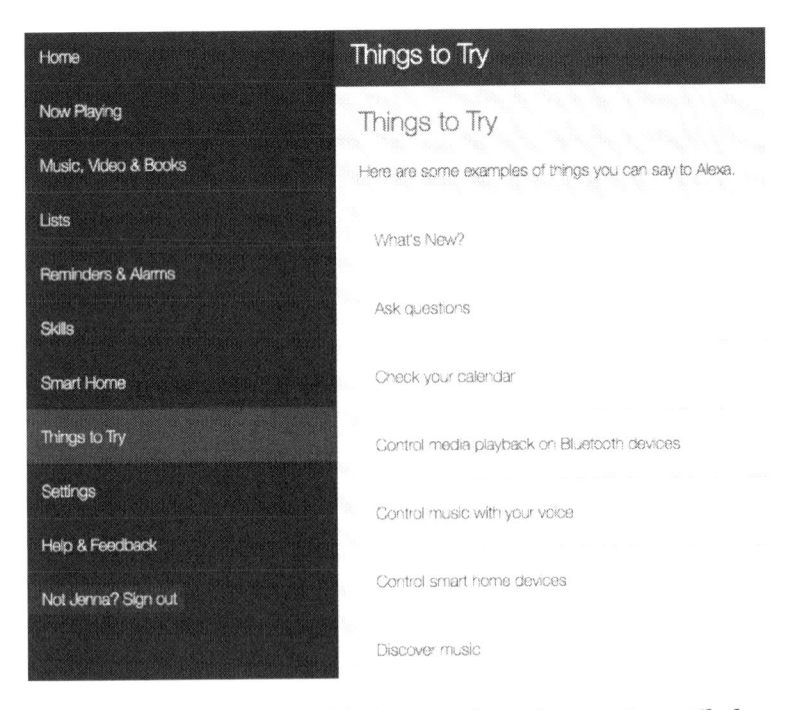

The reason to discuss any of it here rather than wait until the next two chapters is that this list covers the most useful topics, at least in the minds of the folks at Amazon.

We'll cover Help & Feedback and Settings next while trying not to get bogged down in the redundancy built into the app.

In short, we'll cover everything before we're done.

By the way, topics are frequently added and removed from this list, so our discussion that follows might not include all the topics you currently see on the list.

Sign up to our newsletter to stay abreast of developments and changes here - **www.lyntons.com/updates**

Our Approach to Things to Try: We'll list each category in the Things to Try section, give a brief overview and then share tips for getting the most from that Thing to Try. Here we go!

What's New

This is a random and changing assortment of new features and things to try, though Amazon keeps some topics here longer than others.

Some **What's New** topics are explained here. For others, you're invited to ask Alexa about the topics with suggestions like, "*Alexa, Tell me about the all-new Echo/Echo Dot/Echo Spot.*"

Links are included with some topics, usually to other sections on the Things to Try app - and we'll cover those topics as we work through this chapter.

Tips: This section can be a rabbit hole of information, not all of it useful but certainly time-consuming.

We suggest scanning it occasionally, trying a few things that look useful, interesting or fun, and, unless you have a lot of time on your hands, leaving the rest alone.

Echo Dot

This section is all about what Amazon's latest Echo Dot device can do for you.

Tips: Alexa and the Echo Dot gain new capabilities consistently. Reviewing this section for things you didn't know they could do will keep you up to speed and getting maximum value from your investment.

Ask Questions

This topic is aptly named - a list of questions in a spectrum of categories to ask Alexa.

Tips: We're still amazed at how quickly Alexa taps information sources for a wide range of knowledge like the Pittsburgh Pirates won the 1960 World Series and the distance to Jupiter is 391,000,000 miles.

We most often ask how to spell words (well, one of us does), math questions (the other one does that) and general knowledge questions. If you ask a general knowledge question, remember your query will produce a Card viewable on the Home page with a link to a Wikipedia article on the subject.

Calling and Voice Messaging

This is one of the newest features on the Echo devices, Dot-to-Dot, hands-free communication. It's sure to grow with time. We cover this topic in **Chapter 14, Help & Feedback**.

Tips: Read our section on it, first. Then, implement it cautiously. Giving people one more way to get in touch with you can be helpful in some cases and annoying in others. Thankfully, the Dot's Calling feature can be turned off to protect your "quiet time," and turned back on just as easily.

Check Your Calendar

Alexa can link to calendars from Google, Apple and Microsoft. Then, you can ask Alexa questions like, "*How does my day look?*"

Alexa will remind you of your appointments. This feature is discussed in the **Chapter 13, Settings**.

Tips: Couples have the option of having two calendars and linking them, a slightly more complex and hard-to-check setup, or using just one calendar. It really depends on how busy each of you are. We tried both, and that helped us decide which one works best for us, given our circumstances.

Connect Bluetooth Devices

This is a topic covered in **Chapter 2** and again later in **Chapter 14**, Help & Feedback which will be useful if you want to listen to music from your phone or tablet.

Tips: Make sure Bluetooth is turned on in your device's **Settings**. You'll see the Dot listed as **Echo Dot** —*** (where *** is three random letters and numbers) in the list of pairable options.

Control Music

We covered this topic in the Music section (**Chapter 5**), but reading the list on this page of the app might give you an idea or two we missed or you forgot.

Tips: If you have more than one Echo device, you will definitely want to explore the **Multi-Room Music** section covered in Settings, **Chapter 13**.

Control Smart Home Devices

Mostly just a list of things to say, the page does have a useful link at the bottom where you can learn more about Supported Smart Home Devices for Alexa.

Tips: Patience is a virtue! Some devices sync easily; it seems impossible to link others. Review **Chapter 11** on Smart Home, and follow the steps carefully. If they don't work for your device, and you're sure it is compatible, contact the device manufacturer.

Control the Color of Smart Lights

More than 40 smart bulbs now work with Alexa and Dot. Once lighting and Alexa are linked, smart lights provide functionality and customizable ambience you'll enjoy.

Tips: See our tips from the previous section. They apply here too.

Discover Music

Explore what is available in your music options such as Music Library, Prime Music, Spotify and iHeartRadio by asking Alexa questions related to your favorite artists.

Tips: We sometimes get frustrated asking Alexa doe certain music and her not understanding, but saying "***Alexa, what's popular from [artist]?***" always seems to work very well.

Drop In - Alexa's Intercom System

Echo devices can be used like intercoms, a topic covered in the **Calling and Messaging** section of our Help & Feedback chapter.

Tips: There are privacy concerns with Drop In, so read up on the details we provide, and proceed with caution.

Enable Skills

Skills are things you can do with Alexa and Dot with a third-party app, called a Skill.

Tips: Chapter 10 is all about Alexa Skills including plenty of step-by-step guidance. See especially our Tips for Picking Skills to Try and Managing Enabled Skills.

Find Local Businesses and Restaurants

Ask Alexa what restaurants, auto repair shops, clothing stores, dentists, etc., are nearby, and an Alexa will narrate a list.

Now pop back to the Home page of the Alexa app to see the Yelp results displayed in a Card, with helpful details and links.

We cover this in more depth in **Chapter 14**, Help & Feedback.

Tips: Set your device location/address in ***Settings*** under the name of your Dot. Choose the ***Edit*** option, and type in your street (52nd Street), your block (3700 52nd Street) or your complete address (3749 52nd Street). The more exact you are, the more closely the results will be to your location, especially the distance given to each of the businesses.

Find Traffic Information

Alexa will give you the fastest route for your commute or trip and how long it is expected to take.

Tips: Before you're rushed to get out the door to work, or another destination, go to ***Settings*** > ***Traffic*** to enter From and To locations, so all you have to do is ask, "***Alexa, how long is my commute?***"

Fun and Games

This is a list of silly things to interact with Alexa about – Jokes, Alexa's talents, games and more. There's also a link to the Games and Trivia section of Skills (**Chapter 10**).

Tips: Get ready to groan when listening to Alexa's jokes, limericks and puns! They're fun and family friendly, although Alexa's unemotional voice doesn't help comic delivery!

Get Weather Updates

Alexa will give you a weather report based on your location, or you can ask for the temperature or weather in another location. Ask any weather question, and you'll see a 7-day forecast on the **Home** page of your Alexa app.

Tips: For greatest accuracy, edit your Location under ***Settings*** and ***Your Echo Dot***.

Go to the Movies

Alexa narrates a list of movies playing near you as their "posters" appear on the Home page of your Alexa app.

Tap/click a movie for details on story, actors and showtimes. You can also ask for showtimes for a specific film to get details.

Tips: As with many of these Things to Try, giving your location in ***Settings*** will provide results for movies and movie theatres nearest to you.

Hear the News

Your news on the Dot comes in the form of a Flash Briefing. Most news and entertainment organizations produce regular (hourly/

daily/weekly) summaries of local/national/world news or news about their industry.

In the ***Flash Briefing*** section of the **Settings,** you can browse hundreds of content providers. They can be selected one at a time to be added to your Flash Briefing. The order can be edited. Then, when you ask for your Flash Briefing, your news summaries play in the order you've selected.

Tips: Start with content from sources you know and trust, and then expand outward to new sources that sound important, interesting or otherwise appealing.

We have an eclectic mix of about 30 sources. The most important to us, about 7, are first in line, and the rest follow.

With that arrangement, we get the news we want to hear most when time is limited, and we can hear the entire Briefing while we're relaxing on the deck or doing housework.

Keep Up with Your Sports Teams

In ***Settings > Sports Update***, you can search for your favorite teams to add to the list. When you ask for your Sports Update, like a Flash Briefing, you hear the latest scores and upcoming games. Impressively, you can also ask questions about the statistics of any player.

Tips: The **Sports Update** section lists professional leagues in North America and Europe you can select from.

But don't overlook the fact that most college teams are available for your Sports Update too. Just search them.

Listen to Audible Audiobooks

We've covered Audible in detail in **Chapter 6** on Books. This is a service we think is worth trying.

Tips: Try syncing all your connected devices, so you can listen to a book on any of them. Start at home, switch to your phone later, for example, and the book will pick up right where you left off.

Listen to Amazon Music

This feature was thoroughly discussed in the Music section (**Chapter 5**). Amazon Music refers to one of the music subscriptions available from Amazon. About 2 million songs are offered on Prime Music; Music Unlimited has closer to 30 million.

Tips: Try Amazon Prime, not just for the 2 million songs, but many other perks which can be viewed on the Try Prime tab on most pages of **Amazon.com**.

Listen to Kindle Books

Kindle fans will enjoy this feature, though this is listening rather than reading. While Audible books are read by the author or a performer, Kindle books are read by Alexa which isn't so great.

Tips: Certain perks of Kindle Unlimited are part of an Amazon Prime membership. We're not on Amazon's payroll, we just get a lot of value from our Prime membership!

Listen to Podcasts and Radio

This refers to Tunein and iHeartRadio, both covered in the Music section (**Chapter 5**). Both services offer a range of streaming live radio options and archived podcasts.

Tip: Both services have free accounts and premium accounts. We started with a free account and upgraded to a paid option on one of the services. Listening to music, a podcast or a game is so easy with the Dot's hands-free functionality.

Shopping

An Amazon device must feature a shopping option, right? Ask Alexa to shop for an item you want, and options will appear on the Home page to browse, learn more about and purchase, of course.

Tips: If you turn on the Purchase by Voice option in **Settings > Voice Purchasing**, then add a code in the Voice Code option box. This will prevent accidental ordering of stuff you were just browsing.

Set Alarms/Set Reminders/Set Timers

These three sections are very similar, easy to use and have many uses. We covered each in detail in **Chapter 8** where they can be reviewed.

Tips: Review what we said about using Reminders and Timers concurrently when making a large meal. We also use a combination of them during busy days at home when it's easy to forget to tackle tasks we need to complete.

To-dos and Shopping Lists

Creating and managing To-do and Shopping Lists with voice, the Alexa app and the Dot is easy, and the lists are tools you'll use daily, if you're like us.

See our section on Lists (**Chapter 7**) for full details about To-do and Shopping lists.

Tips: Learn both options - voice and the app to create and manage your lists, because there are times, for example, using voice might disturb someone that is sleeping, while at other times, voice is much more convenient than manually interacting with the app. Also, download the Alexa app to your mobile device. Having both lists, not just your shopping list, with you when out and about will be prove helpful.

Use These Phrases Anytime

This list of basic phrases is a good reminder of all the things Alexa and the Dot can do.

Tips: Try some of the suggestions on this list to get more from your device.

More Alexa Products to Try

There's more here than just links to Echo devices. The link to Alexa-enabled products takes you to Amazon where you can browse thermostats, lighting, phones, watches, Bluetooth speakers and many more items that offer built-in Alexa functionality.

Tips: Before you buy an Alexa-enabled device, read lots of reviews, including good and bad, to get a full picture of what the device can and cannot do, its reliability, how hard it is to set up and get connected to Alexa, etc.

13. SETTINGS

Settings is where you turn on or off features and modify others to suit your preferences. In short, Alexa Settings allow you to customize the app and Dot. Let's look together at the Settings page on the Alexa app where we will take the topics from top to bottom (we recommend doing this on a PC or Mac at **alexa.amazon.com**).

Devices

Settings
Devices
Jenna's Echo Dot   Online
Set up a new device
Accounts
Notifications
Music & Media
Flash Briefing
Sports Update
Traffic
Calendar

This section lists by name the devices and apps connected to this **Amazon.com** account and to the accounts of others in your Amazon Household. An option to ***Set Up a New Device*** is given too. When you set up a device, you give it a name. When you download the app to a mobile device or use it online at **alexa.amazon.com**, the apps are named after the device. Whether the device is Online or Offline is shown beneath its name.

Select *Your Echo Dot* from the device list, and the page will change to offer up several options. Let's explore what's there.

Do Not Disturb: Turn this on to stop Calls, Messages, Drop-ins and Notifications. It does not prevent you from interacting with your Dot and using music services, shopping and other features. When you activate *Do Not Disturb* the light ring on top of your Dot flashes purple. Do Not Disturb remains On until you turn it Off.

Scheduled Off: If you have a regular time during the day when you want your Dot to be quiet unless you make a request:

- Choose *Scheduled Off*
- Turn on Scheduled by selecting the *On/Off* box, which will turn blue for On
- Choose *Edit*
- Click or tap the Hour, Minute and AM/PM boxes one at a time, and make your selection for *Daily Start Hour* and *End Hour.* Select *Save Changes* to make the change or select *Cancel* to keep the schedule as it is

Do Not Disturb with Voice: Say, "*Alexa do not disturb*," a chime will sound, Alexa will confirm your request and the screen will darken. The Do Not Disturb icon will be lit in your Dot Settings. Say, "*Alexa, turn off do not disturb*," to reverse these actions.

Sounds: This is where you choose the volume and what sounds you hear from your Dot.

- **Alarm, Timer and Notification Volume:** Click or tap the volume bar, and drag it left for quieter and right for louder. This bar does not change the general volume on the Dot. That can be done by saying, "*Alexa, volume 7*," or using the – and + buttons on top of the Dot.

- **Notifications Sound:** Turn on the box to play a sound when notifications (such as your Amazon order has shipped) or messages (sent by your contacts) arrive. Notifications are explained in **Chapter 8**.

- **Custom Sounds - Alarm:** Currently, custom sounds are available only for the Alarm. Your current sound will be shown; select the box to change the sound. The Alarm sound is discussed in detail in **Chapter 8** on Reminders & Alarms.

- **Request Sounds - Start/End of Request:** If you turn these boxes On, you will hear a sound as soon as Alexa starts listening and when Alexa has stopped listening and is processing your request. Give it a try. We and most of those we've surveyed about it have turned these sounds off.

Device Name: This is the name you gave your Dot at set up. Choose the *Edit* option to change its name.

Device Location: If you haven't set a location, you will see No Zip Code Set with the option to edit that and add a street address. The more specific you make the address, the more accurate the information you'll receive when asking about weather, restaurants, businesses, movie theaters and other nearby locations.

Device Time Zone: Find your country or region in the top *Select a Region* box and your time zone in the *Select a Time Zone* box to get accurate time information.

Wake Word: The default wake word is Alexa. Select the box to change it to Amazon, Echo or Computer and choose *Save*. As we've noted, we tried others and found that Amazon and Computer come up too often in our non-Alexa discussions to use because Alexa wakes but doesn't understand what we're talking about. Echo is a good option, if you don't like using Alexa.

Measurement Units: Change the measuring units to metric by selecting the boxes, or keep/turn them off for imperial units (Fahrenheit, yards, feet, inches, etc.).

About: The only Setting of importance here is the **Reset to Factory Settings** box. If you choose it, your History and Settings modifications will be lost. We suggest only doing this if you sell/give your Dot or if you've set up the Dot on the wrong Amazon account by accident.

Accounts

Do Not Disturb		
Do Not Disturb		On
Scheduled Off		

General

Sounds		
Device name		Jenna's Echo Dot Edit
Device location This location will be used for weather and other local features.		

Go back from your specific Echo Dot settings, to the main **Settings** page to see this **Accounts** section, which is more of a miscellaneous category. Most of the Settings don't require an outside account. We've discussed some of these already. Others we will put off until the next chapter on **Help & Feedback**.

Notifications: There is currently only one notification type here, a *Shipment Notification*. Select it, read the description, and turn on the box if you want to receive a notification from your Dot to make you aware a shipment is expected.

Music & Media: These settings have been covered in the Music section (**Chapter 5**), but here is a brief review. Click any of the *Music Service* tabs for options such as *Link/Unlink your Account* and Go to the service's website. The *Choose Default Music Service* box at the bottom of the list allows you to choose the *Music Library* (Amazon Music or Spotify) and the *Station Service* (Amazon Music, Pandora and iHeartRadio) from where Alexa will get the music you request unless you specify another service in your request. Click or tap your preferences, and choose *Done*.

Flash Briefing: A Flash Briefing is news and information in the form of summaries from hundreds of sources now creating them. This section of Settings allows you to create, review and manage your personalized Flash Briefing:

- The main page shows your current Flash Briefing content providers, whether they are turned On or Off and the order of those turned On.

- Select the ***Get More Flash Briefing Content*** option to see all the sources. Rows are categorized by content and can be scrolled horizontally, and you can scroll down too.

- Once there, another option is to use the Search box to locate specific providers such as a TV/Radio station or network, newspaper, magazine or organization or to see what comes up when you search a genre like entertainment, sports, medicine or politics.

- Most of the sources are Alexa Skills. Review each Skill page for details, or simply Enable the Skill for each source you want as part of your Flash Briefing.

- As you compile a list of Flash Briefing sources, they appear on the main ***Settings > Flash Briefing*** page in the On section with their On/Off boxes blue for On. Select the box to turn any of them Off, and they will move to the lower Off section.

- Select ***Edit Order,*** and the list will be shown with three horizontal lines next to each source. Click or tap and hold the lines for any of them, and drag it up or down in the order.

- To listen, say, "***Alexa, play Flash Briefing***" or simply "***Alexa, Flash Briefing***", and it will begin on your Dot in your preferred order.

- You can control what you hear by saying, ***"Alexa... Skip", Next," Go back"***. You can also say "***Pause***" and "***Resume***"..

Sports Update: This feature is like the Flash Briefing accept that the news is delivered by Alexa:

- Compile the list of teams you want to follow using the Search box. The list of current North American and European professional sports leagues you can select teams from is found on the Things to Try page by selecting **Keep Up with Your Sports Teams**.

- College teams aren't listed, but they can be searched. For example, typing Texas into the search box produces a list of results that includes Texas A&M Aggies Basketball and Texas A&M Aggies Football as separate options along with options for five other colleges.
- Select any team, and it will appear on the main **Sports Update** page in a list showing the team and sport.
- Delete any of them using the **X**.
- You can't drag the teams up or down to change their order. The only option is to delete them and then add them again in the order you want to hear them.
- Say, "**Alexa, Sports Update**" to hear it. Updates are given only for teams that are currently in their season.
- While the update is playing, control what you hear by saying, "**Skip**", "**Next**," "**Go back**"

Traffic: Find out how long your trip will take if you leave now. The Traffic form shows your "From" location as the location you've entered in Settings for your device location: **Settings > Devices > Your device > Device Location**.

Choose **Edit** to change the address. If you give your complete address rather than just your Zip Code or street, your traffic commute time will be more accurate. Select **Settings > Traffic** to:

- View your **From location**, and select **Change Address** if necessary to edit the location
- Add a **To** location or change it with the **Change Address** option. We suggest entering the place you go the most as the To location - your workplace, school, parent's or children's home, for example. Change it anytime.
- Choose the **Add Stop** option to include a location where you plan to stop along the way. This can later be deleted or changed using the options shown.

- Alexa responds to "*Alexa, Traffic update,*" "*My commute,*" "*How's traffic*" and similar requests.
- While Alexa is giving you the update, the drive time is shown along with one or more alternate routes.

Calendar: Alexa works with Google, Microsoft and Apple calendars. Ask Alexa to add an event to your calendar on a day and time, and it will be done.

Alexa will confirm, and your calendar for the next two days will appear on the Home page of your Alexa app. Here's how to set up and use a calendar with Alexa:

- Choose the calendar you want to link.
- Select the *Link your Calendar Account*, and a window will open on the service you've chosen, such as Google.
- Sign into the account. You might have to enter a username, email or phone plus your password.
- You'll be asked to give Alexa permission to have access to your calendar and possibly other information. For Microsoft, Alexa also wants to Read your Profile, Read your Contacts and Access your Info Anytime
- Select *Yes* to link the account or *No* to cancel the link
- Links are shown on the main Calendars page in Settings next to each service. For example, our page says 1 Linked Account next to Microsoft and Google.
- Below the list, you can select the default Calendar from the list. Select the box, and your linked accounts will appear. Choose your preferred Calendar.
- Start adding items to your Calendar using voice. If you leave out a detail such as the day/date or time, Alexa will ask.
- Each added Calendar event is confirmed by Alexa verbally.
- To view or manage your Calendar on your Dot, say, "*Alexa, what's on my Calendar?*" to hear what's there.
- **Note:** You will **NOT receive reminders** from your Dot when an event time arrives. That's why we use **Reminders** in

addition to putting things on the calendar. For example, if we want to be somewhere at 7pm that's a 30-minute drive, we'll set a Reminder - something like "**Remind me at 6:15pm today that it's almost time to leave.**"

- **Note:** If one of your calendars is linked, anyone can say, "*Alexa, what's on my Calendar*" to hear it.
- Select any linked Calendar from the app to manage its Settings such as Show Holidays in the United States
- The Calendar's setting page also has an option to **Unlink this Calendar Account**. Selecting that option unlinks the account with no confirmation or further action required. The **Link your Calendar Account** appears again for future use.

Lists: You have To-do and Shopping Lists with Alexa. You can link those lists to 3rd party list services like **Any.do, AnyList** and **Todoist** and others that have produced Skills.

When you Enable one of these, your Alexa To-do and Shopping Lists will sync with the Skill, and you can use the Skill to read or add items, depending on how you manage the Settings. Your options are available under **Settings > Lists**.

- Select any of the list Skills to view information about the Skill such as a summary of how it works, its rating and individual user reviews.
- Select the blue **Enable** button to try the Skill
- Link your Alexa lists to the Skill service by following the prompts that include giving the Skill provider permission to Read your Alexa Lists and to Write new items onto either list. These permissions can be changed by selecting the **Settings** tab on the Skill. This is also where you can **Unlink** the accounts.
- Once connected, you can go to the Skill provider, **Any.do** for example, to Read or Add items to an Alexa list based on the permissions you've given.
- **Our suggestion:** When it comes to this sort of thing, our motto is "just because it can be done doesn't mean it should be done." If you already use one of these handy list services, then linking the account to Alexa might serve a useful purpose.

We find that with the Alexa app on our computers and mobile devices and available for the asking on our Dot, no other list service is needed. The added convenience of Alexa Reminders makes additional lists unnecessary for us.

Voice Purchasing: To make shopping on Amazon fully hands-free, turn on Voice Purchasing:

- Go to ***Settings > Voice Purchasing***
- Turn on Purchase by Voice by selecting the ***On/Off*** box – it will turn blue
- Select ***View Payment Settings*** to be taken to Amazon.com to ensure you have a valid 1-click payment method on file. The link will open in a new window where you will see ***Payment Settings*** with your ***Default 1-Click Payment Method***. Select ***Edit Payment Method*** to be taken to your payment page on Amazon to manage your payment options.
- **Strongly Advised:** Use the 4-digit Voice Code option to eliminate accidental ordering or unauthorized purchases.
- Your next option is to allow "recognized" speakers to order without the code. Alexa creates what it calls a Voice Profile for each person using its impressive voice recognition technology. If you keep this option Off, as we do, the code will be required for every purchase. If you turn it On, then "recognized speakers" won't have to use a code.

Household Profile: An Amazon Household is a group of Amazonians that share one another's music, books, payment methods and more. Households must contain at least 1 and up to 2 adults, up to 2 teens and up to 4 pre-teens.

To set up a Household go to **www.amazon.com**. Select the ***Accounts & Lists*** tab rather than just hovering over it.

When the page opens, select ***Amazon Household*** in the ***Shopping Programs and Rentals*** box. Alternatively follow this link: **www.amazon.com/myh/manage**.

We've created a Household and enjoy it mostly for sharing music, books and other content.

General

The only Setting of importance here is **History**. Every time you speak to Alexa, a recording is made. Note that this list overlaps with the Cards viewable on the app **Home** page but is not identical. Not all voice interaction with Alexa produces a Card. We don't pay any attention to this, but here's how to manage this History:

- Select any entry to review its details
- Answer whether Alexa heard you properly
- Delete the voice recording
- Note that deleting a voice recording also deletes the Card, but it doesn't work conversely. If you choose **Delete** on a Card, the voice recording remains as part of Alexa's efforts to understand your voice profile more accurately.

To delete all voice recordings at once, you must de-register your Echo Dot, also termed ***Reset to Factory Defaults***. You should deregister/reset your device when getting rid of it.

14. HELP & FEEDBACK

Help & Feedback is the most comprehensive page/section of the app. Many of its topics are listed in **Chapter 12** - Things to Try and explained there.

At the top of the main **Help & Feedback** page is a list of User Guides for Alexa and Echo devices. We'll just cover the Alexa guide as we feel we've already covered anything you're likely to find in the Echo Dot Guide.

Most topics here are explained clearly in the app, so we won't waste your time with rehashing them. These topics may change as Amazon update the Alexa app.

The Alexa App User Guide

Select **Alexa** to open the Guide to browse topics of interest or to find helpful information to make using Alexa easier and more functional.

Alexa Device Support

- **Accessibility Features:** Accessibility functions are built into Alexa, and the app is usable with Vision and Hearing equipment you might already have such as the screen readers listed.
- **About Amazon Alexa:** This is an overview of Alexa/Echo Dot capabilities. While we've covered most topics, if you're still unclear about anything, a wealth of details is here. Let's hit one not covered elsewhere:

Calling and Messaging

Alexa has impressive device-to-device communication skills. Patience in setting them up and learning how to use them will be rewarded. It's not hard, just a multi-step task that is fully detailed in the Alexa app under this heading.

Alexa-to-Alexa Calls: This feature allows you to call another Alexa user on their Echo device or phone or to leave them a message. They must be in your phone's Contacts and be Alexa/Echo device users to be called. Here are the details:

- Download the latest Alexa App for your iOS or Android phone from your app store.

- Select the *Conversations* icon from the mobile app's home page (the middle icon of the three at the bottom of the screen). Follow the prompts to sign up for **Calling & Messaging** and verify your mobile number.

- Import your contacts. The names of other signed-up Alexa/Echo device users will appear, so encourage friends and family who use Alexa to sign up.

- Edit and add contacts in your phone's local address, as normal.

- Open the Alexa app, and ask Alexa to call one of the signed-up contacts.

- You can also call directly from the Alexa app on your phone. Choose the *Conversations* icon and tap a contact. Tap the phone icon for audio-only calls. Tap the camera icon for video calls.

- You can call anyone from your Echo device if they have the Alexa app on their phone, whether they have an Echo or not. The call will go to their phone.

- When receiving a call, Alexa will tell you who is calling and the light ring will be green. Say, "***Alexa answer***", or say, "***Ignore the call***". Say "***Alexa, hang up***" when you've completed the call.

- Messaging is like leaving a voicemail or sending a text. To send a verbal message, say a contacts name. Alexa will guide you through the process.

- To send a voice message from the mobile app, tap the *Conversations* icon, tap *New Conversation* and the contact you want to message. The prompts will walk you through the process. To type a message, at this point tap the ***Keyboard*** icon, type the message and select ***Send***.

- When you receive a message, your Dot's light ring will be yellow. Say, "***Alexa, play my messages***" to hear them. If more than one mobile number is connected with the Alexa account, say, "***Play messages for Tom***," for example, to hear only your messages.

- Reply to messages on the mobile app as you would reply to a text.

- Contact Amazon customer service at 1.877.375.9365 with questions or to cancel Alexa Calling & Messaging.

A few tries with messaging will boost your confidence in using the technology. Once you're familiar with it, hands-free Calling on Alexa iseasy and convenient.

Additional Help: If you want a refresher with visuals, go to - **www.lyntons.com/CandM**

Instructional videos for this and other topics can be found by going to **www.lyntons.com/Avideo**.

In the Alexa app, view all Calling & Messaging topics at ***Help & Feedback > Alexa > Alexa Calling and Messaging***.

Drop In

The Drop In tab and all the setup information needed is found in ***About Amazon Alexa > Make Alexa-to-Alexa Calls and Messages***. Here's an overview of Drop In and the information/links found in that section of the ***Settings > Alexa User Guide***.

The Drop In feature allows you to listen in on other Echo devices in your home, such as when using one as a baby or child monitor, and communicate directly with other Alexa/Echo users that have enabled Drop In without making a call

- **Drop In with same-house Echo devices:** The basic approach to using the **Drop In** feature in your home is to enable Drop In on the Alexa mobile app for each device.

Remember that each Echo device will have its own name such as Jenna's Echo or Tom's Echo Dot. Then, using the app for Jenna's Echo, grant Tom's Echo Dot Drop In permission and vice-versa. Once done, they can drop in on one another by saying "**Alexa drop in on Tom's Echo Dot**," and end the drop in with, "**Alexa hang up**." Drop in can be controlled from the mobile app too.

- **Drop In with friends and family:** The approach is about the same. You use the Alexa mobile app and the Conversations icon at the bottom to select **Contacts** from your list and grant them Drop In permission. Both parties must enable Drop In and grant the other permission.

Privacy concerns are obvious. There are three levels of Permission:

- **On** - All contacts you have granted permission to can drop in anytime on your device
- **Only my household** - Only devices that are registered on your Alexa account can drop in
- **Off** - Drop in can be turned off on any device

Step-by-step instructions are buried in Settings at: **Help & Feedback > Alexa User Guide > About Calling & Messaging > Drop In**. There, you will find several sub-pages that will guide you through Use/Permissions/Remove Contacts and more.

Music, Video and Book

We haven't completely exhausted this topic. There are a few more things worth covering:

- **Play Music on Multiple Echo Devices:** This is a Smart Home feature that is easy to set up in six quick steps outlined in this section on the app: **Help & Feedback > Alexa User Guide > Music, Video & Books > Play Music on Multiple Echo Devices**. Form and Name the group as shown, and it will appear in the **Smart Home** page of your Alexa app under Groups. Then, request the music you want to hear and control it with voice or the app the same way you would on one device. The one downside is that external Bluetooth speakers and other services like Audible and Flash Briefing don't yet work with this feature.

- **Set Up Alexa Connected Speaker Skills:** In this section, step-by-step directions are given to set up a Sonos or other external speaker to play your entire range of music. Find and enable the speaker's Skill in the Skill tab, sign in and link accounts, follow the onscreen prompts including logging into your Amazon account. Once that's set up, go to ***Smart Home > Devices > Discover***. The two will sync, and you'll be playing music on the external speaker. Use the suggested requests given on the speaker's Skills page.

Tip: If you haven't bought an external speaker yet, find the Skills page for those you're considering or search Speakers on the Skills tab. Read ratings and reviews about how the speaker works with Alexa. They'll provide insight that will produce a better purchasing decision.

News, Weather & Traffic

The last topic in this New, Weather & Traffic section, **Ask Alexa,** contains many helpful requests that will broaden your appreciation for all Alexa can do and all the information it can deliver on the spot.

Smart Home

There's lots of step-by-step guidance in this section of the **Help & Feedback** section to support what we discussed in **Chapter 11**. Let's look at a couple:

- **Remove a Smart Home Device:** If you want to remove a smart device from Alexa for any reason – it's not working and it's causing interference with other devices are common reasons - these four easy steps will get it done. You might also want to remove/forget the device if it needs to be reset or if you're having trouble with the initial connection.

- **Supported Colors and Shades of White for Alexa Smart Light Control:** Smart lights are also discussed in **Chapter 11**, but this list gives you all your color options when selecting colors and shades for supported smart bulbs.

Shopping

Alexa is a handy shopping tool, just what you'd expect from Amazon.

- **About Placing Orders on Amazon:** This topic is about setting up an Amazon Prime account, something we've covered and recommended, and the purchasing Prime-eligible items and Amazon's Choice items (highly rated, well-priced products, according to Amazon). If you don't have Prime, you can use Alexa to add items to your shopping cart on the website and to track order status. Finally of note, Primenow is Amazon's courier service for fast delivery of products and restaurant carryout. See **Primenow.amazon.com** for details.

- **Manage Voice Purchasing Options:** We're being a bit redundant here since this was covered in the Accounts Settings, but it's a significant feature of Alexa and Echo Dot. If you have a Prime account, you can set up hands-free voice purchases, a feature we enjoy. You have to enable Voice Purchasing in **Settings > Accounts**. There, you can also add a 4-digit code that will be required before a voice sale will go through. We recommend the code to prevent accidental and unauthorized purchases.

- **Order an Alexa Device with our Voice:** Just one of your voice purchasing options, you can buy up to 12 Echo devices and up to 24 Echo Dots. Go for it! The holidays, birthdays, Father's Day, Mother's Day, Just Because...great gift ideas!

- **Track Open Orders with Alexa:** Very handy and as easy as asking, "*Alexa, where is my stuff?*" You can also turn on Notifications in **Settings > Notifications > Shopping Notifications** to read and/or hear notifications about when an Amazon order is out for delivery or has been left at your door.

- **Reorder from Amazon Restaurants with Alexa:** If you have a Prime membership, you can order meals from restaurants in your delivery area and then reorder the same thing with Alexa. Your default payment method will be used, and the meal will be delivered to your default address. The Amazon Restaurant Skill must be enabled. Currently it's hovering at 2.5 Stars out of 5, so you might want to review the ratings when you're reading this and read recent reviews to see if the service has improved. You can find all participating restaurants in your area at **Primenow.Amazon.com/restaurants**.
- **Cancel your Order:** If you have buyer's regret in the few minutes after placing an order, say, "*Alexa, cancel my order.*" If it was done in time, Alexa confirms that the order was cancelled. The rest of this section is about cancelling orders on the Amazon website if it's too late for Alexa to do it.
- **Alexa and Alexa Device FAQs:** Can't find answers anywhere else? Try here. Since the Help & Feedback section is so large that navigation can be frustrating, we'll give you the path to the FAQs: *Help & Feedback > Alexa > Shopping > Alexa and Alexa Device FAQs*. That's also an odd spot for it, in our opinion.

Alexa Quick Fixes

These topics have been covered "here and there" in this guide, but it is nice to have them all in one place. Here is how to find the section in the Alexa app and what topics are available: Go to *Help & Feedback > Alexa > Alexa Quick Fixes* for:

- *Alexa App doesn't work*
- *Alexa doesn't discover Your Smart Home device*
- *Quick fixes for using Alexa with Smart Home cameras*
- *Alexa doesn't understand you*
- *Bluetooth issues with Alexa devices*
- *Multi-room music quick fixes*
- *Streaming issues on Alexa devices*
- *Quick fixes for Alexa Skills*

There are a couple more items in the Help & Feedback page worth mentioning. They are:

Contact Us

If you have stubborn problems with the Alexa app or Echo Dot, Amazon Customer Service is helpful and timely. Email and phone give you options for your preferred way to communicate.

E-mail Customer Service: Select a device you have questions about, select a category closest and specific issue, and tell Amazon all about it.

Be Called by Customer Service: Use the same form to choose a device, category and issue, and then leave your phone number before selecting the best time to be called.

Send Feedback: Send Amazon your thoughts on Alexa and the Echo Dot. Be sure to select the box if you want Amazon to follow-up by contacting you.

Printed in Great Britain
by Amazon